Post Quantum Cryptography

Post Quantum Cryptography

Ella Long

Kruger Brentt
Publishers
2026

Kruger Brentt Publishers UK. LTD.
Company Number 9728962

Regd. Office: 68 St Margarets Road, Edgware, Middlesex HA8 9UU

ISBN: 978-1-78715-624-1

For information on all our publications visit our website at http://krugerbrentt.com/

Preface

The rapid advancement of quantum computing poses a profound challenge to traditional cryptographic systems, as many widely used encryption algorithms could be rendered vulnerable in the near future. Post-quantum cryptography (PQC) addresses this emerging threat by developing cryptographic schemes that are secure against attacks from both classical and quantum computers. This field is crucial for ensuring the confidentiality, integrity, and authenticity of data in the era of quantum technology.

This book, *Post Quantum Cryptography*, provides a comprehensive introduction to the principles, algorithms, and practical applications of cryptography designed to resist quantum attacks. It is intended for students, researchers, cybersecurity professionals, and practitioners in information technology and computer science who seek a deep understanding of secure communication in a post-quantum world.

The opening chapters introduce the fundamentals of cryptography, including classical encryption, public-key infrastructure, and computational complexity. Subsequent sections delve into the vulnerabilities posed by quantum computing and the mathematical foundations of post-quantum algorithms, including lattice-based, code-based, multivariate, and hash-based cryptography. The book explains the design principles, security assumptions, and performance considerations of these emerging cryptographic schemes.

Practical applications of post-quantum cryptography, including secure communications, digital signatures, key exchange protocols, and blockchain technology, are discussed with illustrative examples. The book also examines standardization efforts, implementation challenges, and best practices for transitioning existing systems to quantum-resistant solutions.

By combining theoretical foundations, algorithmic insights, and practical perspectives, Post Quantum Cryptography equips readers with the knowledge required to understand, implement, and evaluate secure cryptographic systems in the era of quantum computing.

Ella Long

Preface

Contents

CHAPTER 1

Introduction to Post-Quantum Cryptography

1.1 WHAT IS POST-QUANTUM CRYPTOGRAPHY?

Post-quantum cryptography (PQC), sometimes referred to as quantum-proof, quantum-safe, or quantum-resistant, is the development of cryptographic algorithms (usually public-key algorithms) that are expected (though not confirmed) to be secure against a cryptanalytic attack by a quantum computer. Most widely used public-key algorithms rely on the difficulty of one of three mathematical problems: the integer factorization problem, the discrete logarithm problem or the elliptic-curve discrete logarithm problem. All of these problems could be easily solved on a sufficiently powerful quantum computer running Shor's algorithm or possibly alternatives.

As of 2024, quantum computers lack the processing power to break widely used cryptographic algorithms; however, because of the length of time required for migration to quantum-safe cryptography, cryptographers are already designing new algorithms to prepare for Y2Q or Q-Day, the day when current algorithms will be vulnerable to quantum computing attacks. Mosca's theorem provides the risk analysis framework that helps organizations identify how quickly they need to start migrating.

Their work has gained attention from academics and industry through the PQCrypto conference series hosted since 2006, several workshops on Quantum Safe Cryptography hosted by the European Telecommunications Standards Institute (ETSI), and the Institute for Quantum Computing. The rumoured existence of widespread harvest now, decrypt later programs has also been seen as a motivation for the early introduction of post-quantum algorithms, as data recorded now may still remain sensitive many years into the future.

In contrast to the threat quantum computing poses to current public-key algorithms, most current symmetric cryptographic algorithms and hash functions are considered to be relatively secure against attacks by quantum computers. While the quantum Grover's algorithm does speed up attacks against symmetric ciphers, doubling the key size can effectively counteract these attacks. Thus post-quantum symmetric cryptography does not need to differ significantly from current symmetric cryptography. In 2024, the U.S. National Institute of Standards and Technology (NIST) released final versions of its first three Post-Quantum Cryptography Standards.

1.1.1 Preparation

Digital infrastructures require robust cybersecurity. Cryptographic systems are vital to protect the confidentiality and authenticity of data. Quantum computing will be a threat to many of the cryptographic algorithms used to achieve these protection goals. Data that is currently not quantum-safe, whether it is stored or transmitted, and that must remain confidential for a long time, may be compromised in the future by quantum computers ("store now, decrypt later" attacks). In addition, authenticity will also be jeopardised by quantum computers. The threat that quantum computing poses to cybersecurity can be countered by a timely, comprehensive and coordinated transition to post-quantum cryptography (PQC).

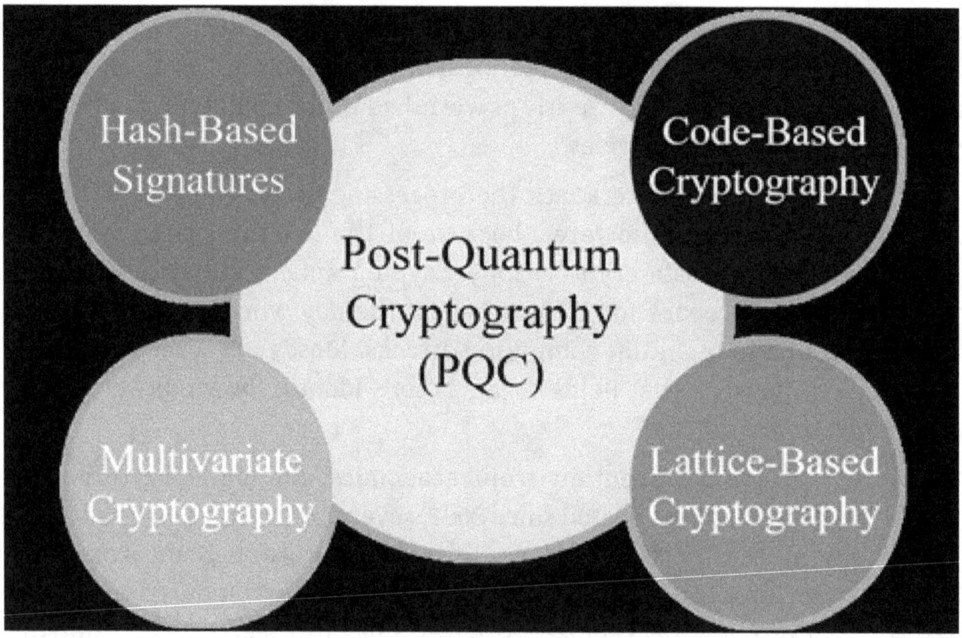

Fig: Basic types of Post-Quantum Cryptography (PQC)

1.1.2 Algorithms

- **Lattice-Based Cryptography:** This approach includes cryptographic systems such as learning with errors, ring learning with errors (ring-LWE), the ring learning with errors key exchange and the ring learning with errors signature, the older NTRU or GGH encryption schemes, and the newer NTRU signature and BLISS signatures. Some of these schemes like NTRU encryption have been studied for many years without anyone finding a feasible attack. Others like the ring-LWE algorithms have proofs that their security reduces to a worst-case problem. The Post-Quantum Cryptography Study Group sponsored by the European Commission suggested that the Stehle–Steinfeld variant of NTRU be studied for standardization rather than the NTRU algorithm. At that time, NTRU was still patented. Studies have indicated that NTRU may have more secure properties than other lattice based algorithms.

- **Multivariate Cryptography:** This includes cryptographic systems such as the Rainbow (Unbalanced Oil and Vinegar) scheme which is based on the difficulty of solving systems of multivariate equations. Various attempts to build secure multivariate equation encryption schemes have failed. However, multivariate signature schemes like Rainbow could provide the basis for a quantum secure digital signature. The Rainbow Signature Scheme is patented (the patent expires in August 2029).
- **Hash-Based Cryptography:** This includes cryptographic systems such as Lamport signatures, the Merkle signature scheme, the XMSS, the SPHINCS, and the WOTS schemes. Hash based digital signatures were invented in the late 1970s by Ralph Merkle and have been studied ever since as an interesting alternative to number-theoretic digital signatures like RSA and DSA. Their primary drawback is that for any hash-based public key, there is a limit on the number of signatures that can be signed using the corresponding set of private keys. This fact reduced interest in these signatures until interest was revived due to the desire for cryptography that was resistant to attack by quantum computers. There appear to be no patents on the Merkle signature scheme and there exist many non-patented hash functions that could be used with these schemes. The stateful hash-based signature scheme XMSS developed by a team of researchers under the direction of Johannes Buchmann is described in RFC 8391. Moni Naor and Moti Yung invented UOWHF hashing in 1989 and designed a signature based on hashing (the Naor-Yung scheme) which can be unlimited-time in use (the first such signature that does not require trapdoor properties).
- **Code-Based Cryptography:** This includes cryptographic systems which rely on error-correcting codes, such as the McEliece and Niederreiter encryption algorithms and the related Courtois, Finiasz and Sendrier Signature scheme. The original McEliece signature using random Goppa codes has withstood scrutiny for over 40 years. However, many variants of the McEliece scheme, which seek to introduce more structure into

the code used in order to reduce the size of the keys, have been shown to be insecure. The Post-Quantum Cryptography Study Group sponsored by the European Commission has recommended the McEliece public key encryption system as a candidate for long term protection against attacks by quantum computers.

- **Isogeny-Based Cryptography:** These cryptographic systems rely on the properties of isogeny graphs of elliptic curves (and higher-dimensional abelian varieties) over finite fields, in particular supersingular isogeny graphs, to create cryptographic systems. Among the more well-known representatives of this field are the Diffie–Hellman-like key exchange CSIDH, which can serve as a straightforward quantum-resistant replacement for the Diffie–Hellman and elliptic curve Diffie–Hellman key-exchange methods that are in widespread use today, and the signature scheme SQIsign which is based on the categorical equivalence between supersingular elliptic curves and maximal orders in particular types of quaternion algebras. Another widely noticed construction, SIDH/SIKE, was spectacularly broken in 2022. The attack is however specific to the SIDH/SIKE family of schemes and does not generalize to other isogeny-based constructions.
- **Symmetric Key Quantum Resistance:** Provided one uses sufficiently large key sizes, the symmetric key cryptographic systems like AES and SNOW 3G are already resistant to attack by a quantum computer. Further, key management systems and protocols that use symmetric key cryptography instead of public key cryptography like Kerberos and the 3GPP Mobile Network Authentication Structure are also inherently secure against attack by a quantum computer. Given its widespread deployment in the world already, some researchers recommend expanded use of Kerberos-like symmetric key management as an efficient way to get post-quantum cryptography today.

1.1.3 Security Reductions

In cryptography research, it is desirable to prove the equivalence of a cryptographic algorithm and a known hard mathematical problem. These proofs are often called "security reductions", and are used to demonstrate the difficulty of cracking the encryption algorithm. In other words, the security of a given cryptographic algorithm is reduced to the security of a known hard problem. Researchers are actively looking for security reductions in the prospects for post-quantum cryptography.

- **Lattice-Based Cryptography:** Ring-LWE Signature: In some versions of Ring-LWE there is a security reduction to the shortest-vector problem

(SVP) in a lattice as a lower bound on the security. The SVP is known to be NP-hard. Specific ring-LWE systems that have provable security reductions include a variant of Lyubashevsky's ring-LWE signatures defined in a paper by Güneysu, Lyubashevsky, and Pöppelmann. The GLYPH signature scheme is a variant of the Güneysu, Lyubashevsky, and Pöppelmann (GLP) signature which takes into account research results that have come after the publication of the GLP signature in 2012. Another Ring-LWE signature is Ring-TESLA. There also exists a "derandomized variant" of LWE, called Learning with Rounding (LWR), which yields "improved speedup (by eliminating sampling small errors from a Gaussian-like distribution with deterministic errors) and bandwidth". While LWE utilizes the addition of a small error to conceal the lower bits, LWR utilizes rounding for the same purpose.

- **Lattice-Based Cryptography:** NTRU, BLISS: The security of the NTRU encryption scheme and the BLISS signature is believed to be related to, but not provably reducible to, the closest vector problem (CVP) in a lattice. The CVP is known to be NP-hard. The Post-Quantum Cryptography Study Group sponsored by the European Commission suggested that the Stehle–Steinfeld variant of NTRU, which does have a security reduction be studied for long term use instead of the original NTRU algorithm.
- **Multivariate Cryptography:** Unbalanced Oil and Vinegar: Unbalanced Oil and Vinegar signature schemes are asymmetric cryptographic primitives based on multivariate polynomials over a finite field $\mathbb{F}$. Bulygin, Petzoldt and Buchmann have shown a reduction of generic multivariate quadratic UOV systems to the NP-Hard multivariate quadratic equation solving problem.
- **Hash-Based Cryptography:** Merkle Signature Scheme: In 2005, Luis Garcia proved that there was a security reduction of Merkle Hash Tree signatures to the security of the underlying hash function. Garcia showed in his paper that if computationally one-way hash functions exist then the Merkle Hash Tree signature is provably secure. Therefore, if one used a hash function with a provable reduction of security to a known hard problem one would have a provable security reduction of the Merkle tree signature to that known hard problem. The Post-Quantum Cryptography Study Group sponsored by the European Commission has recommended use of Merkle signature scheme for long term security protection against quantum computers.
- **Code-Based Cryptography:** McEliece: The McEliece Encryption System has a security reduction to the syndrome decoding problem (SDP). The

SDP is known to be NP-hard. The Post-Quantum Cryptography Study Group sponsored by the European Commission has recommended the use of this cryptography for long term protection against attack by a quantum computer.

- **Code-Based Cryptography:** RLCE: In 2016, Wang proposed a random linear code encryption scheme RLCE which is based on McEliece schemes. RLCE scheme can be constructed using any linear code such as Reed-Solomon code by inserting random columns in the underlying linear code generator matrix.
- **Supersingular Elliptic Curve Isogeny Cryptography:** Security is related to the problem of constructing an isogeny between two supersingular curves with the same number of points. The most recent investigation of the difficulty of this problem is by Delfs and Galbraith indicates that this problem is as hard as the inventors of the key exchange suggest that it is. There is no security reduction to a known NP-hard problem.

1.1.4 Comparison

One common characteristic of many post-quantum cryptography algorithms is that they require larger key sizes than commonly used "pre-quantum" public key algorithms. There are often tradeoffs to be made in key size, computational efficiency and ciphertext or signature size. The table lists some values for different schemes at a 128-bit post-quantum security level.

Algorithm	Type	Public key	Private key	Signature
ML-DSA	Lattice	1,312 B	2,560 B	2,420 B
NTRU Encrypt	Lattice	766.25 B	842.875 B	
Streamlined NTRU Prime	Lattice	154 B		
Rainbow	Multivariate	124 kB	95 kB	
SPHINCS	Hash Signature	1 kB	1 kB	41 kB
SPHINCS+	Hash Signature	32 B	64 B	8 kB
BLISS-II	Lattice	7 kB	2 kB	5 kB
GLP-Variant GLYPH Signature	Ring-LWE	2 kB	0.4 kB	1.8 kB
NewHope	Ring-LWE	2 kB	2 kB	
Goppa-based McEliece	Code-based	1 MB	11.5 kB	
Random Linear Code based encryption	RLCE	115 kB	3 kB	
Quasi-cyclic MDPC-based McEliece	Code-based	1,232 B	2,464 B	

SIDH	Isogeny	564 B	48 B	
SIDH (compressed keys)	Isogeny	330 B	48 B	
3072-bit Discrete Log	not PQC	384 B	32 B	96 B
256-bit Elliptic Curve	not PQC	32 B	32 B	65 B

A practical consideration on a choice among post-quantum cryptographic algorithms is the effort required to send public keys over the internet. From this point of view, the Ring-LWE, NTRU, and SIDH algorithms provide key sizes conveniently under 1 kB, hash-signature public keys come in under 5 kB, and MDPC-based McEliece takes about 1 kB. On the other hand, Rainbow schemes require about 125 kB and Goppa-based McEliece requires a nearly 1 MB key.

- **Lattice-Based Cryptography:** LWE Key Exchange and Ring-LWE Key Exchange: The fundamental idea of using LWE and Ring LWE for key exchange was proposed and filed at the University of Cincinnati in 2011 by Jintai Ding. The basic idea comes from the associativity of matrix multiplications, and the errors are used to provide the security. The paper appeared in 2012 after a provisional patent application was filed in 2012. In 2014, Peikert presented a key transport scheme following the same basic idea of Ding's, where the new idea of sending additional 1 bit signal for rounding in Ding's construction is also utilized. For somewhat greater than 128 bits of security, Singh presents a set of parameters which have 6956-bit public keys for the Peikert's scheme. The corresponding private key would be roughly 14,000 bits. In 2015, an authenticated key exchange with provable forward security following the same basic idea of Ding's was presented at Eurocrypt 2015, which is an extension of the HMQV construction in Crypto2005. The parameters for different security levels from 80 bits to 350 bits, along with the corresponding key sizes are provided in the paper.
- **Lattice-Based Cryptography:** NTRU Encryption: For 128 bits of security in NTRU, Hirschhorn, Hoffstein, Howgrave-Graham and Whyte, recommend using a public key represented as a degree 613 polynomial with coefficients mod (2 10) $\bmod(2^{10})$. This results in a public key of 6130 bits. The corresponding private key would be 6743 bits.
- **Multivariate Cryptography:** Rainbow Signature: For 128 bits of security and the smallest signature size in a Rainbow multivariate quadratic equation signature scheme, Petzoldt, Bulygin and Buchmann, recommend using equations in GF(31) with a public key size of just over 991,000 bits, a private key of just over 740,000 bits and digital signatures which are 424 bits in length.

- **Hash-Based Cryptography:** Merkle Signature Scheme: In order to get 128 bits of security for hash based signatures to sign 1 million messages using the fractal Merkle tree method of Naor Shenhav and Wool the public and private key sizes are roughly 36,000 bits in length.
- **Code-Based Cryptography:** McEliece: For 128 bits of security in a McEliece scheme, The European Commission's Post-Quantum Cryptography Study group recommends using a binary Goppa code of length at least $n = 6960$ and dimension at least $k = 5413$, and capable of correcting $t = 119$ errors. With these parameters the public key for the McEliece system will be a systematic generator matrix whose non-identity part takes $k \times (n - k) = 8373911$ bits. The corresponding private key, which consists of the code support with $n = 6960$ elements from $GF(2^{13})$ and a generator polynomial of with $t = 119$ coefficients from $GF(2^{13})$, will be 92,027 bits in length. The group is also investigating the use of Quasi-cyclic MDPC codes of length at least $n = 2^{16} + 6 = 65542$ and dimension at least $k = 2^{15} + 3 = 32771$, and capable of correcting $t = 264$ errors. With these parameters the public key for the McEliece system will be the first row of a systematic generator matrix whose non-identity part takes $k = 32771$ bits. The private key, a quasi-cyclic parity-check matrix with $d = 274$ nonzero entries on a column (or twice as much on a row), takes no more than $d \times 16 = 4384$ bits when represented as the coordinates of the nonzero entries on the first row. Barreto et al. recommend using a binary Goppa code of length at least $n = 3307$ and dimension at least $k = 2515$, and capable of correcting $t = 66$ errors. With these parameters the public key for the McEliece system will be a systematic generator matrix whose non-identity part takes $k \times (n - k) = 1991880$ bits. The corresponding private key, which consists of the code support with $n = 3307$ elements from $GF(2^{12})$ and a generator polynomial of with $t = 66$ coefficients from $GF(2^{12})$, will be 40,476 bits in length.
- **Supersingular Elliptic Curve Isogeny Cryptography:** For 128 bits of security in the supersingular isogeny Diffie–Hellman (SIDH) method, De Feo, Jao and Plut recommend using a supersingular curve modulo a 768-bit prime. If one uses elliptic curve point compression the public key will need to be no more than 8x768 or 6144 bits in length. A March 2016 paper by authors Azarderakhsh, Jao, Kalach, Koziel, and Leonardi showed how to cut the number of bits transmitted in half, which was further improved by authors Costello, Jao, Longa, Naehrig, Renes and Urbanik resulting in a compressed-key version of the SIDH protocol with public keys only 2640 bits in size. This makes the number of bits

transmitted roughly equivalent to the non-quantum secure RSA and Diffie–Hellman at the same classical security level.

- **Symmetric-Key–Based Cryptography:** As a general rule, for 128 bits of security in a symmetric-key–based system, one can safely use key sizes of 256 bits. The best quantum attack against arbitrary symmetric-key systems is an application of Grover's algorithm, which requires work proportional to the square root of the size of the key space. To transmit an encrypted key to a device that possesses the symmetric key necessary to decrypt that key requires roughly 256 bits as well. It is clear that symmetric-key systems offer the smallest key sizes for post-quantum cryptography.

1.1.5 Forward Secrecy

A public-key system demonstrates a property referred to as perfect forward secrecy when it generates random public keys per session for the purposes of key agreement. This means that the compromise of one message cannot lead to the compromise of others, and also that there is not a single secret value which can lead to the compromise of multiple messages. Security experts recommend using cryptographic algorithms that support forward secrecy over those that do not. The reason for this is that forward secrecy can protect against the compromise of long term private keys associated with public/private key pairs. This is viewed as a means of preventing mass surveillance by intelligence agencies. Both the Ring-LWE key exchange and supersingular isogeny Diffie–Hellman (SIDH) key exchange can support forward secrecy in one exchange with the other party. Both the Ring-LWE and SIDH can also be used without forward secrecy by creating a variant of the classic ElGamal encryption variant of Diffie–Hellman.

1.1.6 Open Quantum Safe Project

The Open Quantum Safe (OQS) project was started in late 2016 and has the goal of developing and prototyping quantum-resistant cryptography. It aims to integrate current post-quantum schemes in one library: liboqs. liboqs is an open source C library for quantum-resistant cryptographic algorithms. It initially focuses on key exchange algorithms but by now includes several signature schemes. It provides a common API suitable for post-quantum key exchange algorithms, and will collect together various implementations. liboqs will also include a test harness and benchmarking routines to compare performance of post-quantum implementations. Furthermore, OQS also provides integration of liboqs into OpenSSL. As of March 2023, the following key exchange algorithms are supported: As of August 2024, NIST has published 3 algorithms below as FIPS standards and the 4th is expected near end of the year:

Algorithm	Type
FIPS-203: CRYSTALS-Kyber	ML-KEM: Module Learning With Error
Classic McEliece	goppa codes
BIKE	codes
HQC	codes
Frodo	Learning with errors
NTRU	Lattice-based cryptography
FIPS-204: CRYSTALS-Dilithium	ML-DSA: Module Short Integer Solution
FIPS-206: Falcon	FN-DSA: Short Integer Solution
FIPS-205: SPHINCS+	SLH-DSA: hash based

Older supported versions that have been removed because of the progression of the NIST Post-Quantum Cryptography Standardization Project are:

Algorithm	Type
BCNS15	Ring learning with errors key exchange
NewHope	Ring learning with errors key exchange
SIDH	Supersingular isogeny key exchange
McBits	Error-correcting codes

1.1.7 Implementation

One of the main challenges in post-quantum cryptography is considered to be the implementation of potentially quantum safe algorithms into existing systems. There are tests done, for example by Microsoft Research implementing PICNIC in a PKI using Hardware security modules. Test implementations for Google's NewHope algorithm have also been done by HSM vendors. In August 2023, Google released a FIDO2 security key implementation of an ECC/Dilithium hybrid signature schema which was done in partnership with ETH Zürich. The Signal Protocol uses Post-Quantum Extended Diffie–Hellman (PQXDH).

On February 21, 2024, Apple announced that they were going to upgrade their iMessage protocol with a new PQC protocol called "PQ3", which will utilize ongoing keying. Apple stated that, although quantum computers don't exist yet, they wanted to mitigate risks from future quantum computers as well as so-called "Harvest now, decrypt later" attack scenarios. Apple stated that they believe their PQ3 implementation provides protections that "surpass those in all other widely deployed messaging apps", because it utilizes ongoing keying. Apple intends to fully replace the existing iMessage protocol within all supported conversations with PQ3 by the end of 2024. Apple also defined a scale to make it easier to compare the security properties of messaging apps, with a scale represented by levels ranging from 0 to 3: 0 for no end-to-end by default, 1 for pre-quantum end-to-end by default, 2 for PQC key establishment only (e.g. PQXDH), and 3 for PQC key establishment and ongoing rekeying (PQ3).

Other notable implementations include:

- bouncycastle
- liboqs

1.2 THE QUANTUM THREAT TO MODERN CRYPTOGRAPHY

Modern cryptography is the backbone of cybersecurity, protecting everything from personal messages to critical infrastructure. It employs mathematical techniques to secure data – ensuring confidentiality, integrity, and authenticity of information in transit and storage. Every day, encryption shields countless digital interactions: securing email and me ssaging, safeguarding online banking and e-commerce transactions, and protecting state secrets. In fact, encryption carries a heavy load in modern digitized so ciety, guarding electronic secrets like email content, medical records, and infor mation vital to national security. Thanks to cryptography, sensitive data can traverse public networks unreadable to anyone but the intended recipient making it an indispensable tool in the cyber defender's arsenal.

At the heart of many cryptographic systems is the concept of a one-way function – a mathematical operation that's easy to perform in one direction but extremely difficult to reverse without a special key. One prominent example is the multiplication of large prime numbers: multiplying two primes is easy, but finding the original prime factors from their product (a process called factorization) is incredibly hard. Modern encryption algorithms leverage this asymmetry. RSA, one of the most widely used public-key cryptosystems, bases its security on the difficulty of factoring large composite numbers. The public-key in RSA includes a very large number N (hundreds or even thousands of bits long) that is the product of two secret prime numbers. The security of RSA "relies on the practical difficulty of factoring the product of two large prime numbers" (Wikipedia) – an assumption that has held true against classical computers for decades.

Because factoring large numbers is so computationally onerous, RSA and similar cryptosystems have been trusted to secure digital communications worldwide. Until now, breaking RSA by factorizing its key has been effectively impossible with classical computing power. For instance, factoring an 829-bit, 250-digit RSA number (known as RSA-250) using state-of-the-art classical methods took about 2,700 CPU-years of effort (Wikipedia). To put that in perspective, RSA keys used in practice are typically 2048 bits or more – vastly larger than 829 bits – and would require astronomically mor e time to cra ck by brute force. In practical terms, a classical computer might take longer than the age of the universe to factor a 2048-bit RSA key using known algorithms. This enormous gap between the ease of multiplying primes and the infeasibility of factoring the product underpins the security of modern encryption.

However, a new kind of computing threatens to upend this delicate balance. Quantum computing, leveraging principles of quantum physics, promises to solve certain problems much faster than classical machines. In the 1990s, mathematician Peter Shor discovered a quantum algorithm that could factor large numbers exponentially faster than any known classical algorithm . Shor's Algorithm demonstrated theoretically that a sufficiently advanced quantum computer could crack RSA – and other cryptosystems based on large-numb er factorization or related problems – in a feasible amount of time. This looming capability has put the cybersecurity community on alert. As researchers race to build more powerful quantum computers, the question is no longer if our current encryption schemes can be broken, but when.

We will begin with a refresher on RSA encryption and why factoring is fundamental to its security. Then we'll explore how quantum computing differs from classical computing – introducing key concepts like superposition and entanglement – and why these quantum properties enable algorithms like Shor's to succeed where classical methods fail. We will break down Shor's Algorithm conceptually, step by step, using real-world analogies to demystify how it factors large numbers so efficiently. We will examine the profound cybersecurity implications: how a future quantum computer running Shor's Algorithm could threaten RSA, ECC (elliptic curve cryptography), and most public-key systems in use today, and how close we are to that reality. Finally, we will discuss post-quantum cryptography (PQC) – the emerging suite of new cryptographic algorithms designed to resist quantum attacks – and outline strategies for transitioning to these quantum-safe methods. Governments, standards bodies, and industry experts are already urging proactive migration to PQC, and we will summarize their recommendations so that security professionals can prepare for the coming quantum era.

1.3 OVERVIEW OF QUANTUM ALGORITHMS (SHOR'S, GROVER'S)

Grover's algorithm, also known as the quantum search algorithm, is a quantum algorithm for unstructured search that finds with high probability the unique input to a black box function that produces a particular output value, using just $O(\sqrt{N})$ evaluations of the function, where N is the size of the function's domain. It was devised by Lov Grover in 1996. The analogous problem in classical computation would have a query complexity $O(N)$ (i.e., the function would have to be evaluated $O(N)$ times: there is no better approach than trying out all input values one after the other, which, on average, takes $N/2$ steps).

Charles H. Bennett, Ethan Bernstein, Gilles Brassard, and Umesh Vazirani proved that any quantum solution to the problem needs to evaluate the function $\Omega(\sqrt{N})$ times, so Grover's algorithm is asymptotically optimal. Since classical algorithms for NP-complete problems require exponentially many steps, and Grover's algorithm provides at most a quadratic speedup over the classical solution for unstructured search, this suggests that Grover's algorithm by itself will not provide polynomial-time solutions for NP-complete problems (as the square root of an exponential function is still an exponential, not a polynomial function).

Unlike other quantum algorithms, which may provide exponential speedup over their classical counterparts, Grover's algorithm provides only a quadratic speedup. However, even quadratic speedup is considerable when N is large, and Grover's algorithm can be applied to speed up broad classes of algorithms. Grover's algorithm could brute-force a 128-bit symmetric cryptographic key in roughly 2^{64} iterations, or a 256-bit key in roughly 2^{128} iterations. It may not be the case that Grover's algorithm poses a significantly increased risk to encryption over existing classical algorithms, however.

1.3.1 Applications and Limitations

Grover's algorithm, along with variants like amplitude amplification, can be used to speed up a broad range of algorithms. In particular, algorithms for NP-complete problems which contain exhaustive search as a subroutine can be sped up by Grover's algorithm. The current theoretical best algorithm, in terms of worst-case complexity, for 3SAT is one such example. Generic constraint satisfaction problems also see quadratic speedups with Grover.

These algorithms do not require that the input be given in the form of an oracle, since Grover's algorithm is being applied with an explicit function, e.g. the function checking that a set of bits satisfies a 3SAT instance. However, it is unclear whether Grover's algorithm could speed up best practical algorithms for these problems. Grover's algorithm can also give provable speedups for black-box problems in quantum query complexity, including element distinctness and the collision problem (solved with the Brassard–Høyer–Tapp algorithm). In these types of problems, one treats the oracle function f as a database, and the goal is to use the quantum query to this function as few times as possible.

Cryptography

Grover's algorithm essentially solves the task of function inversion. Roughly speaking, if we have a function $y = f(x)$ that can be evaluated on a quantum computer, Grover's algorithm allows us to calculate x x when given y y. Consequently, Grover's algorithm gives broad asymptotic speed-ups to many kinds of brute-force attacks on symmetric-key cryptography, including collision attacks and pre-image attacks. However, this may not necessarily be the most efficient algorithm since, for example, the Pollard's rho algorithm is able to find a collision in SHA-2 more efficiently than Grover's algorithm.

Limitations

Grover's original paper described the algorithm as a database search algorithm, and this description is still common. The database in this analogy is a table of all of the function's outputs, indexed by the corresponding input. However, this database is not represented explicitly. Instead, an oracle is invoked to evaluate an item by its index. Reading a full database item by item and converting it into such a representation may take a lot longer than Grover's search. To account for such effects, Grover's algorithm can be viewed as solving an equation or satisfying a constraint. In such applications, the oracle is a way to check the constraint and is not related to the search algorithm. This separation usually prevents algorithmic optimizations, whereas conventional search algorithms often rely on such optimizations and avoid exhaustive search. Fortunately, fast Grover's oracle implementation is possible for many constraint satisfaction and optimization problems. The major barrier to instantiating a speedup from Grover's algorithm is that the quadratic speedup achieved is too modest to overcome the large overhead of near-term quantum computers. However, later generations of fault-tolerant quantum computers with better hardware performance may be able to realize these speedups for practical instances of data.

1.3.2 Problem Description

As input for Grover's algorithm, suppose we have a function $f:\{0,1,\ldots,N-1\}\to\{0,1\}$. In the "unstructured database" analogy, the domain represent indices to a database, and f(x) = 1 if and only if the data that x points to satisfies the search criterion. We additionally assume that only one index satisfies f(x) = 1, and we call this index ω. Our goal is to identify ω.

We can access f with a subroutine (sometimes called an oracle) in the form of a unitary operator U_ω that acts as follows:

$$\begin{cases} U_\omega |x\rangle = -|x\rangle & \text{for } x=\omega, \text{ that is, } f(x)=1, \\ U_\omega |x\rangle = |x\rangle & \text{for } x\neq\omega, \text{ that is, } f(x)=0. \end{cases}$$

This uses the N N-dimensional state space $\mathcal{H}$, which is supplied by a register with $n=[\log_2 N]$ qubits. This is often written as

$$U_\omega |x\rangle = (-1)^{f(x)} |x\rangle.$$

Grover's algorithm outputs ω with probability at least 1/2 using $O(\sqrt{N})$ applications of U_ω. This probability can be made arbitrarily large by running Grover's algorithm multiple times. If one runs Grover's algorithm until ω is found, the expected number of applications is still $O(\sqrt{N})$, since it will only be run twice on average.

Alternative Oracle Definition

This stage compares the oracle U_ω with an oracle U_f.

U_ω is different from the standard quantum oracle for a function f. This standard oracle, denoted here as U_f, uses an ancillary qubit system. The operation then represents an inversion (NOT gate) on the main system conditioned by the value of f(x) from the ancillary system:

$$\begin{cases} U_f |x\rangle|y\rangle = |x\rangle|\neg y\rangle & \text{for } x=\omega, \text{ that is, } f(x)=1, \\ U_f |x\rangle|y\rangle = |x\rangle|y\rangle & \text{for } x\neq\omega, \text{ that is, } f(x)=0, \end{cases}$$

Or briefly,

$$U_f |x\rangle|y\rangle = |x\rangle|y\oplus f(x)\rangle.$$

These oracles are typically realized using uncomputation.

If we are given U_f as our oracle, then we can also implement U_ω, since U_ω is U_f when the ancillary qubit is in the state $|-\rangle = \frac{1}{\sqrt{2}}\big(|0\rangle - |1\rangle\big) = H|1\rangle$:

$$\begin{aligned} U_f\big(|x\rangle\otimes|-\rangle\big) &= \frac{1}{\sqrt{2}}\big(U_f|x\rangle|0\rangle - U_f|x\rangle|1\rangle\big) \\ &= \frac{1}{\sqrt{2}}(|x\rangle|0\oplus f(x)\rangle - |x\rangle|1\oplus f(x)\rangle) \\ &= \begin{cases} \frac{1}{\sqrt{2}}(-|x\rangle|0\rangle + |x\rangle|1\rangle) & \text{if } f(x)=1, \\ \frac{1}{\sqrt{2}}(|x\rangle|0\rangle - |x\rangle|1\rangle) & \text{if } f(x)=0 \end{cases} \\ &= (U_\omega|x\rangle)\otimes|-\rangle \end{aligned}$$

So, Grover's algorithm can be run regardless of which oracle is given. If U_f is given, then we must maintain an additional qubit in the state $|-\rangle$ and apply U_f in place of U_ω.

1.3.3 Algorithm

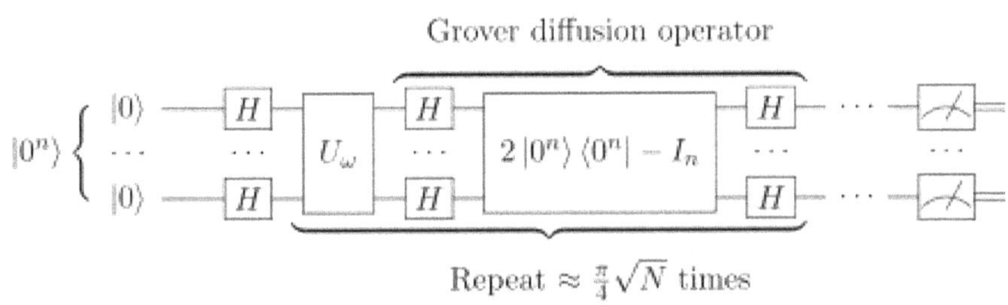

Fig: Quantum circuit representation of Grover's algorithm

The steps of Grover's algorithm are given as follows:

- Initialize the system to the uniform superposition over all states: $|s\rangle = \frac{1}{\sqrt{N}}\sum_{x=0}^{N-1}|x\rangle$.
- Perform the following "Grover iteration" $r(N)$ times:
  - Apply the operator U_ω
  - Apply the Grover diffusion operator $U_s = 2|s\rangle\langle s| - I$
- Measure the resulting quantum state in the computational basis.

For the correctly chosen value of r, the output will be $|\omega\rangle$ with probability approaching 1 for N ≫ 1. Analysis shows that this eventual value for $r(N)$ satisfies $r(N) \leq \left\lceil \frac{\pi}{4}\sqrt{N} \right\rceil$.

Implementing the steps for this algorithm can be done using a number of gates linear in the number of qubits. Thus, the gate complexity of this algorithm is $O(\log(N) r(N))$, or $O(\log(N))$ per iteration.

1.3.4 Geometric Proof

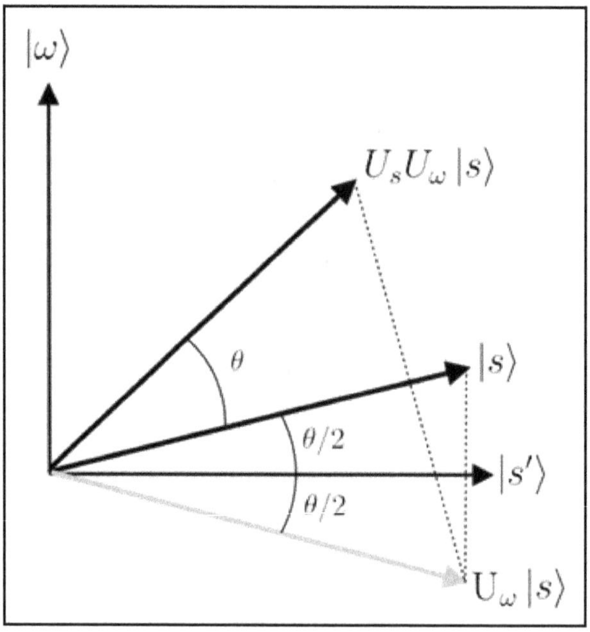

Fig: Picture Showing the Geometric Interpretation of the First Iteration of Grover's Algorithm

There is a geometric interpretation of Grover's algorithm, following from the observation that the quantum state of Grover's algorithm stays in a two-dimensional subspace after each step. Consider the plane spanned by $|s\rangle$ and $|\omega\rangle$; equivalently, the plane spanned by $|\omega\rangle$ and the perpendicular ket $|s'\rangle = \frac{1}{\sqrt{N-1}}\sum_{x\neq\omega}|x\rangle$.

Grover's algorithm begins with the initial ket $|s\rangle$, which lies in the subspace. The operator U_ω is a reflection at the hyperplane orthogonal to $|\omega\rangle$ for vectors in the plane spanned by $|s'\rangle$ and $|\omega\rangle$, i.e. it acts as a reflection across $|s'\rangle$. This can be seen by writing U_ω in the form of a Householder reflection:

$$U_\omega = I - 2|\omega\rangle\langle\omega|.$$

The operator $U_s = 2|s\rangle\langle s| - I$ is a reflection through $|s\rangle$. Both operators U_s and $|s'\rangle$ take states in the plane spanned by $|s'\rangle$ and $|\omega\rangle$ to states in the plane. Therefore, Grover's algorithm stays in this plane for the entire algorithm.

It is straightforward to check that the operator $U_s U_\omega$ of each Grover iteration step rotates the state vector by an angle of $\theta = 2\arcsin\frac{1}{\sqrt{N}}$. So, with enough iterations, one can rotate from the initial state $|s'\rangle$ to the desired output state $|\omega\rangle$. The initial ket is close to the state orthogonal to $|\omega\rangle$:

$$\langle s'|s\rangle = \sqrt{\frac{N-1}{N}}.$$

In geometric terms, the angle $\theta/2$ between $|s\rangle$ and $|s'\rangle$ is given by

$$\sin\frac{\theta}{2} = \frac{1}{\sqrt{N}}.$$

We need to stop when the state vector passes close to $|\omega\rangle$; after this, subsequent iterations rotate the state vector away from $|\omega\rangle$, reducing the probability of obtaining the correct answer. The exact probability of measuring the correct answer is

$$\sin^2\left(\left(r+\frac{1}{2}\right)\theta\right),$$

where r is the (integer) number of Grover iterations. The earliest time that we get a near-optimal measurement is therefore $r \approx \pi\sqrt{N}/4$.

1.3.5 Algebraic Proof

To complete the algebraic analysis, we need to find out what happens when we repeatedly apply $U_s U_\omega$. A natural way to do this is by eigenvalue analysis of a matrix. Notice that during the entire computation, the state of the algorithm is a linear combination of $s \; and \; \omega$. We can write the action of $U_s \; and \, U_\omega$ in the space spanned by $\{|s\rangle, |\omega\rangle\}$ as:

$$U_s : a|\omega\rangle + b|s\rangle \mapsto [|\omega\rangle |s\rangle]\begin{bmatrix} -1 & 0 \\ 2/\sqrt{N} & 1 \end{bmatrix}\begin{bmatrix} a \\ b \end{bmatrix}.$$

$$U_\omega : a|\omega\rangle + b|s\rangle \mapsto [|\omega\rangle |s\rangle]\begin{bmatrix} -1 & -2/\sqrt{N} \\ 0 & 1 \end{bmatrix}\begin{bmatrix} a \\ b \end{bmatrix}.$$

So in the basis $\{|\omega\rangle, |s\rangle\}$ (which is neither orthogonal nor a basis of the whole space) the action $U_s U_\omega$ of applying U_ω followed by U_s is given by the matrix

$$U_sU_\omega = \begin{bmatrix} -1 & 0 \\ 2/\sqrt{N} & 1 \end{bmatrix}\begin{bmatrix} -1 & -2/\sqrt{N} \\ 0 & 1 \end{bmatrix} = \begin{bmatrix} 1 & 2/\sqrt{N} \\ -2/\sqrt{N} & 1-4/N \end{bmatrix}.$$

This matrix happens to have a very convenient Jordan form. If we define $t = \arcsin(1/\sqrt{N})$, it is

$$U_sU_\omega = M\begin{bmatrix} e^{2it} & 0 \\ 0 & e^{-2it} \end{bmatrix}M^{-1}$$

$$where\ M = \begin{bmatrix} -i & i \\ e^{it} & e^{-it} \end{bmatrix}.$$

It follows that r-th power of the matrix (corresponding to r iterations) is

$$(U_sU_\omega)^r = M\begin{bmatrix} e^{2rit} & 0 \\ 0 & e^{-2rit} \end{bmatrix}M^{-1}.$$

Using this form, we can use trigonometric identities to compute the probability of observing ω after r iterations mentioned in the previous stage,

$$\left|[\langle\omega|\omega\rangle \quad \langle\omega|s\rangle](U_sU_\omega)^r\begin{bmatrix} 0 \\ 1 \end{bmatrix}\right|^2 = \sin^2\left((2r+1)t\right).$$

Alternatively, one might reasonably imagine that a near-optimal time to distinguish would be when the angles 2rt and -2rt are as far apart as possible, which corresponds to $2rt \approx \pi/2$, or $r = \pi/4t = \pi/4\arcsin(1/\sqrt{N}) \approx \pi\sqrt{N}/4$. Then the system is in state

$$[|\omega\rangle|s\rangle](U_sU_\omega)^r\begin{bmatrix} 0 \\ 1 \end{bmatrix} \approx [|\omega\rangle|s\rangle]M\begin{bmatrix} i & 0 \\ 0 & -i \end{bmatrix}M^{-1}\begin{bmatrix} 0 \\ 1 \end{bmatrix} = |\omega\rangle\frac{1}{\cos(t)} - |s\rangle\frac{\sin(t)}{\cos(t)}.$$

A short calculation now shows that the observation yields the correct answer ω with error $O\left(\frac{1}{N}\right)$.

1.3.6 Extensions and Variants

Multiple Matching Entries

If, instead of 1 matching entry, there are k matching entries, the same algorithm works, but the number of iterations must be $\frac{\pi}{4}\left(\frac{N}{k}\right)^{1/2}$ instead of $\frac{\pi}{4}N^{1/2}$.

There are several ways to handle the case if k is unknown. A simple solution performs optimally up to a constant factor: run Grover's algorithm repeatedly for increasingly small values of k, e.g., taking k = N, N/2, N/4, ..., and so on, taking $k = N/2^t$ for iteration t until a matching entry is found. With sufficiently high probability, a marked entry will be found by iteration $t = \log_2(N/k) + c$ for some constant c. Thus, the total number of iterations taken is at most

$$\frac{\pi}{4}\left(1+\sqrt{2}+\sqrt{4}+\cdots+\sqrt{\frac{N}{k2^{c}}}\right)=O\left(\sqrt{N/k}\right).$$

Another approach if k is unknown is to derive it via the quantum counting algorithm prior. If $k = N/2$ (or the traditional one marked state Grover's Algorithm if run with $N = 2$), the algorithm will provide no amplification. If $k > N/2$, increasing k will begin to increase the number of iterations necessary to obtain a solution. On the other hand, if $k \geq N/2$, a classical running of the checking oracle on a single random choice of input will more likely than not give a correct solution.

A version of this algorithm is used in order to solve the collision problem.

Quantum Partial Search

A modification of Grover's algorithm called quantum partial search was described by Grover and Radhakrishnan in 2004. In partial search, one is not interested in finding the exact address of the target item, only the first few digits of the address. Equivalently, we can think of "chunking" the search space into blocks, and then asking "in which block is the target item?". In many applications, such a search yields enough information if the target address contains the information wanted. For instance, to use the example given by L. K. Grover, if one has a list of students organized by class rank, we may only be interested in whether a student is in the lower 25%, 25–50%, 50–75% or 75–100% percentile. To describe partial search, we consider a database separated into K blocks, each of size $b = N/K$. The partial search problem is easier. Consider the approach we would take classically - we pick one block at random, and then perform a normal search through the rest of the blocks (in set theory language, the complement). If we do not find the target, then we know it is in the block we did not search. The average number of iterations drops from $N/2\, to\, (N-b)/2.$

Grover's algorithm requires $\frac{\pi}{4}\sqrt{N}$ iterations. Partial search will be faster by a numerical factor that depends on the number of blocks K. Partial search uses n_1 global iterations and n_2 local iterations. The global Grover operator is designated G_1 and the local Grover operator is designated G_2. The global Grover operator acts on the blocks. Essentially, it is given as follows:

- Perform j_1 standard Grover iterations on the entire database.
- Perform j_2 local Grover iterations. A local Grover iteration is a direct sum of Grover iterations over each block.
- Perform one standard Grover iteration.

The optimal values of $j_1\, and\, j_2$ are discussed in the paper by Grover and Radhakrishnan. One might also wonder what happens if one applies successive partial searches at different levels of "resolution". This idea was studied in detail by Vladimir Korepin and Xu, who called it binary quantum search. They proved that it is not in fact any faster than performing a single partial search.

1.3.7 Optimality

Grover's algorithm is optimal up to sub-constant factors. That is, any algorithm that accesses the database only by using the operator U_ω must apply U_ω at least a $1-o(1)$ fraction as many times as Grover's algorithm. The extension of Grover's algorithm to k matching entries, $\pi(N/k)^{1/2}/4$, is also optimal. This result is important in understanding the limits of quantum computation.

If the Grover's search problem was solvable with $\log^c N$ applications of U_ω, that would imply that NP is contained in BQP, by transforming problems in NP into Grover-type search problems. The optimality of Grover's algorithm suggests that quantum computers cannot solve NP-Complete problems in polynomial time, and thus NP is not contained in BQP.

It has been shown that a class of non-local hidden variable quantum computers could implement a search of an N N-item database in at most $O(\sqrt[3]{N})$ steps. This is faster than the $O(\sqrt{N})$ steps taken by Grover's algorithm.

1.4 TIMELINE AND URGENCY: WHY PREPARE NOW?

The timeline for quantum computing's impact on cryptography remains uncertain, but the risks of inaction are clear. Organizations that delay preparing for PQC expose themselves to significant vulnerabilities, including the potential compromise of long-term encrypted data, regulatory penalties, and reputational damage. The transition to quantum-resistant cryptography is a complex and time-consuming process that requires immediate attention to ensure a smooth and secure migration.

Given the rapid advancements in quantum technology, the window for preparation is closing. Governments, businesses, and individuals must recognize the urgency of this shift and begin implementing PQC solutions today. The cost of waiting until quantum computers are operational is far greater than the investment required to secure data now. By taking proactive steps, organizations can safeguard their information, maintain compliance, and secure their position in an increasingly quantum-vulnerable world. The future of cryptography is being written today, and those who act now will be best prepared for the challenges ahead.

1.4.1 Quantum Threat: A Looming Reality

Quantum computing's ability to break classical encryption stems from algorithms like Shor's algorithm, which can efficiently factor large numbers and solve discrete logarithms—problems that form the basis of modern public-key cryptography. While current quantum computers are still in their infancy, with error rates and qubit counts far below what is needed to crack encryption, progress in the field is accelerating. Companies like IBM, Google, and startups such as Rigetti and IonQ are making steady advancements, and governments worldwide are investing heavily in quantum research.

The exact timeline for when quantum computers will pose a real threat remains debated. Some estimates suggest that within the next 10 to 20 years, quantum machines could reach the necessary scale to break widely used encryption standards. However, the risk is not limited to future data—it also applies to information encrypted today. Attackers could already be harvesting encrypted data with the intention of decrypting it once quantum computers become capable. This "harvest now, decrypt later" strategy means that sensitive information, such as financial records, medical data, and classified communications, could be compromised retroactively if not secured with quantum-resistant algorithms.

1.4.2 Long Lifespan of Encrypted Data

One of the most compelling reasons to act now is the long-term nature of encrypted data. Many industries rely on encryption to protect information that must remain secure for decades. For example, healthcare providers store patient records that must be kept confidential for extended periods, financial institutions maintain transaction histories for regulatory compliance, and governments archive classified intelligence that must remain protected indefinitely. If this data is encrypted using current standards, it could be vulnerable to future quantum attacks, even if those attacks are not yet possible. The implications of this vulnerability are profound. A single breach of long-term encrypted data could have catastrophic consequences, including financial fraud, identity theft, and national security risks. Organizations that fail to transition to PQC in time may find themselves facing legal liabilities, regulatory penalties, and irreversible reputational damage. The cost of inaction far outweighs the investment required to implement quantum-resistant cryptography today.

1.4.3 Complexity of Transitioning to Post-Quantum Cryptography

Transitioning to PQC is not a simple software update but a comprehensive overhaul of cryptographic infrastructure. The process involves multiple stages, including risk assessment, algorithm selection, testing, and deployment across various systems and devices. Unlike previous cryptographic upgrades, PQC requires entirely new mathematical foundations, meaning that existing protocols must be redesigned or replaced. This complexity necessitates early preparation to avoid disruptions in security and operations. Standardization efforts, such as those led by the National Institute of Standards and Technology (NIST), are critical in this transition. NIST has been evaluating PQC algorithms for years and has already selected several candidates for standardization. However, even after final standards are published, organizations will need time to integrate these new algorithms into their systems. Early adopters will have a significant advantage, as they can conduct thorough testing and ensure compatibility before quantum threats materialize.

1.4.4 Regulatory and Compliance Pressures

As the quantum threat becomes more apparent, governments and regulatory bodies are likely to impose stricter requirements for PQC adoption. Industries handling sensitive data, such as finance, healthcare, and defense, will face increasing pressure to demonstrate compliance with quantum-resistant security standards. Organizations that delay preparation may find themselves scrambling to meet new regulations, potentially facing fines, legal consequences, and operational disruptions. Beyond compliance, there is also the matter of maintaining customer and stakeholder trust. In an era where data breaches frequently make headlines, organizations that proactively adopt PQC will be seen as leaders in cybersecurity. Conversely, those that lag behind may suffer reputational harm, losing the confidence of clients, partners, and investors. The shift to PQC is not just a technical necessity but a strategic business decision that can enhance an organization's credibility and resilience.

1.4.5 Competitive Advantage of Early Preparation

While the transition to PQC presents challenges, it also offers opportunities for organizations that act swiftly. Early adopters can position themselves as industry leaders in cybersecurity, gaining a competitive edge in markets where data protection is paramount. Companies that invest in PQC research and implementation today will be better equipped to handle future threats, ensuring business continuity and reducing the risk of costly security breaches. Moreover, early preparation allows organizations to influence the development of PQC standards and best practices. By participating in industry collaborations and contributing to standardization efforts, businesses can shape the future of cryptographic security in a way that aligns with their operational needs. This proactive approach not only enhances security but also fosters innovation and collaboration within the cybersecurity community.

1.5 STAKEHOLDERS: GOVERNMENT, INDUSTRY, ACADEMIA

The threat posed by quantum computing to modern cryptography is real and imminent, requiring urgent action from governments, industries, and academia. Each stakeholder has a critical role to play—governments must lead in standardization and policy, industries must adopt quantum-resistant solutions, and academia must drive research and education. The consequences of inaction are severe, ranging from national security risks to financial instability and loss of public trust. By collaborating and acting now, these stakeholders can ensure a smooth and secure transition to post-quantum cryptography, safeguarding data and infrastructure for the quantum era. The time to prepare is not when quantum computers arrive—it is today.

1.5.1 Role of Governments

Governments hold a unique and critical responsibility in the transition to PQC, as they oversee national security, regulatory frameworks, and public infrastructure. The potential for quantum computers to break classical encryption poses a direct threat to state secrets, military communications, and critical infrastructure such as power grids and financial systems. A failure to prepare could result in espionage, cyber warfare, and economic sabotage on an unprecedented scale. Recognizing this, several nations have already begun investing in quantum research and PQC standardization efforts.

The United States, through the National Institute of Standards and Technology (NIST), has taken a leading role in evaluating and standardizing PQC algorithms. Other countries, including China, the European Union, and Russia, are also accelerating their quantum initiatives, understanding that cryptographic superiority will be a key determinant of global influence in the coming decades. Governments must not only fund research but also mandate the adoption of PQC in critical sectors, ensuring that public and private entities comply with quantum-resistant security measures. Additionally, international cooperation will be essential to prevent a fragmented cryptographic landscape where some nations remain vulnerable while others advance.

Beyond standardization, governments must also prepare for the geopolitical implications of quantum computing. The ability to decrypt adversaries' communications could shift the balance of power, making early adoption of PQC a matter of national security. Intelligence agencies, defense departments, and cybersecurity organizations must collaborate to develop strategies that protect classified information while also preparing for potential quantum-based cyber threats. The time to act is now, as the transition to PQC will take years, and the consequences of delay could be irreversible.

1.5.2 Industry's Responsibility

Industries across sectors—finance, healthcare, technology, and telecommunications—rely on encryption to protect sensitive data, intellectual property, and customer information. The rise of quantum computing threatens to undermine these protections, making PQC adoption not just a technical upgrade but a business imperative. Companies that fail to transition risk financial losses, legal liabilities, and reputational damage from data breaches. The stakes are particularly high for industries that handle long-term data, such as banking, where transaction records must remain secure for decades. The financial sector, for instance, depends on encryption to secure online transactions, digital signatures, and customer authentication. If quantum computers can break these protections, fraud and cyber theft could surge, destabilizing global markets. Similarly, healthcare providers store vast amounts of patient data that must remain confidential under regulations like HIPAA. A quantum attack could expose this information, leading to identity theft, medical fraud, and legal consequences. Technology companies, meanwhile, must ensure that their cloud services, software, and hardware are quantum-resistant to maintain customer trust and competitive advantage.

The transition to PQC is not without challenges. Many organizations still rely on legacy systems that may not support new cryptographic algorithms, requiring significant upgrades. Additionally, the integration of PQC must be seamless to avoid disruptions in operations. Despite these hurdles, early adopters will gain a strategic edge, positioning themselves as leaders in cybersecurity and innovation. Companies that invest in PQC research, collaborate with governments and academia, and begin testing quantum-resistant solutions today will be better prepared for the future. Delaying action could result in costly breaches, regulatory penalties, and loss of market share to more forward-thinking competitors.

1.5.3 Academia's Role

Academic institutions serve as the foundation for research and development in PQC, driving the theoretical and practical innovations necessary for a secure transition. Universities and research organizations are at the forefront of discovering new cryptographic algorithms, analyzing their security, and optimizing them for real-world applications. Without academic contributions, the development of PQC would stall, leaving governments and industries without the tools needed to defend against quantum threats. One of the primary challenges in PQC is ensuring that new algorithms are not only quantum-resistant but also efficient and scalable. Researchers must explore various mathematical approaches, such as lattice-based, hash-based, and code-based cryptography, to identify the most secure and practical solutions. Academia also plays a crucial role in stress-testing these algorithms, identifying vulnerabilities, and refining them before standardization. Collaborations between universities, government agencies, and private companies are essential to accelerate progress and ensure that PQC solutions are robust and widely adopted.

Beyond research, academia is responsible for educating the next generation of cryptographers, cybersecurity experts, and quantum scientists. As the demand for PQC expertise grows, universities must expand their curricula to include quantum computing and post-quantum cryptography. Workforce development is critical, as the shortage of skilled professionals could hinder the global transition to PQC.

By fostering talent and innovation, academic institutions will help bridge the gap between theoretical advancements and practical implementation, ensuring that society is prepared for the quantum era.

1.5.4 Urgency of Collaboration among Stakeholders

The transition to PQC is not a task that any single stakeholder can accomplish alone. Governments, industries, and academia must work together to develop, standardize, and deploy quantum-resistant cryptography before it is too late. Governments provide the regulatory and financial support needed for research and adoption, industries implement and test real-world solutions, and academia drives innovation and education. Without this collaboration, the global transition to PQC will be fragmented, inefficient, and potentially ineffective.

One of the most pressing concerns is the "harvest now, decrypt later" strategy, where adversaries collect encrypted data today with the intention of decrypting it once quantum computers are capable. This threat underscores the need for immediate action, as data secured with classical encryption could already be at risk. Stakeholders must prioritize the migration to PQC, ensuring that new systems are in place before quantum attacks become feasible. Public-private partnerships, international cooperation, and open research initiatives will be key to accelerating progress and minimizing vulnerabilities.

CHAPTER 2

Cryptographic Foundations

2.1 CLASSICAL CRYPTOGRAPHY PRIMER

2.1.1 Public-Key

Public-key cryptography, or asymmetric cryptography, is the field of cryptographic systems that use pairs of related keys. Each key pair consists of a public key and a corresponding private key. Key pairs are generated with cryptographic algorithms based on mathematical problems termed one-way functions. Security of public-key cryptography depends on keeping the private key secret; the public key can be openly distributed without compromising security. There are many kinds of public-key cryptosystems, with different security goals, including digital signature, Diffie–Hellman key exchange, public-key key encapsulation, and public-key encryption.

Public key algorithms are fundamental security primitives in modern cryptosystems, including applications and protocols that offer assurance of the confidentiality and authenticity of electronic communications and data storage. They underpin numerous Internet standards, such as Transport Layer Security (TLS), SSH, S/MIME, and PGP. Compared to symmetric cryptography, public-key cryptography can be too slow for many purposes, so these protocols often combine symmetric cryptography with public-key cryptography in hybrid cryptosystems.

Before the mid-1970s, all cipher systems used symmetric key algorithms, in which the same cryptographic key is used with the underlying algorithm by both the sender and the recipient, who must both keep it secret. Of necessity, the key in every such system had to be exchanged between the communicating parties in some secure way prior to any use of the system – for instance, via a secure channel. This requirement is never trivial and very rapidly becomes unmanageable as the number of participants increases, or when secure channels are not available, or when, (as is sensible cryptographic practice), keys are frequently changed. In particular, if messages are meant to be secure from other users, a separate key is required for each possible pair of users.

By contrast, in a public-key cryptosystem, the public keys can be disseminated widely and openly, and only the corresponding private keys need be kept secret.

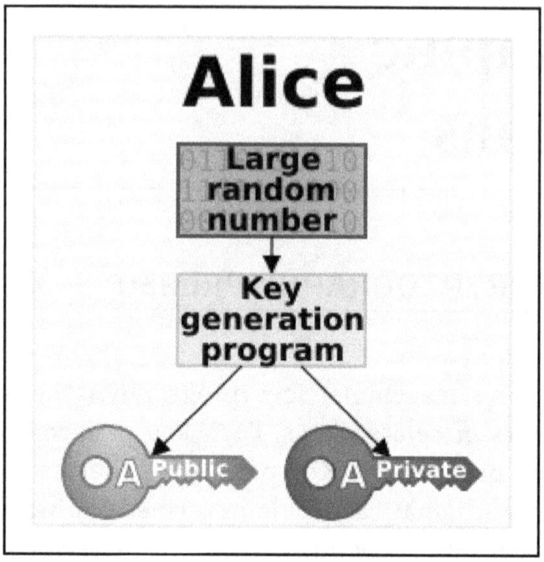

Fig: An Unpredictable (Typically Large and Random) Number is Used to Begin Generation of an Acceptable Pair of Keys Suitable for Use by an Asymmetric Key Algorithm

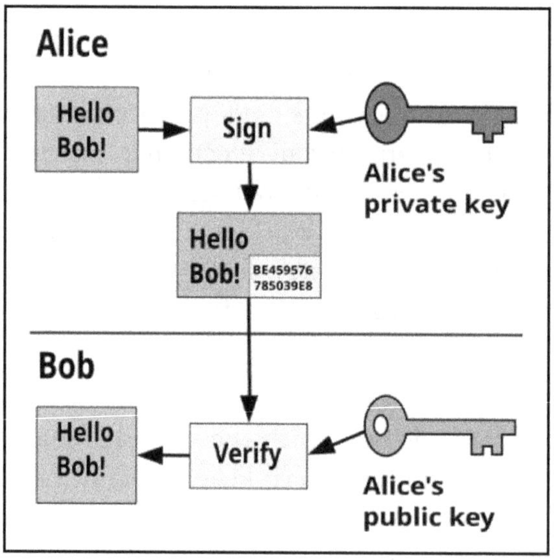

Fig: In this Example the Message is Digitally Signed with Alice's Private Key, but the Message itself is not Encrypted. 1) Alice Signs a Message with her Private Key. 2) Using Alice's Public Key, Bob can Verify that Alice Sent the Message and that the Message has not been Modified

The two best-known types of public key cryptography are digital signature and public-key encryption:

- In a digital signature system, a sender can use a private key together with a message to create a signature. Anyone with the corresponding public key can verify whether the signature matches the message, but a forger who does not know the private key cannot find any message/ signature pair that will pass verification with the public key. For example, a software publisher can create a signature key pair and include the public key in software installed on computers. Later, the publisher can distribute an update to the software signed using the private key, and any computer receiving an update can confirm it is genuine by verifying the signature using the public key. As long as the software publisher keeps the private key secret, even if a forger can distribute malicious updates to computers, they cannot convince the computers that any malicious updates are genuine.

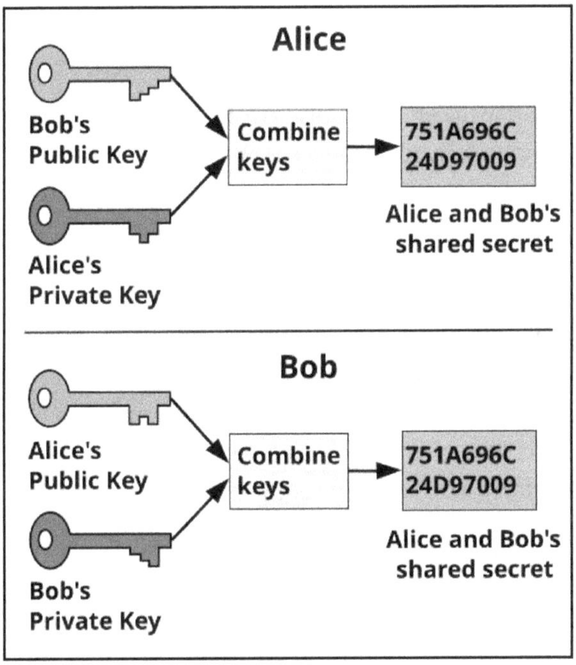

Fig: In the Diffie-Hellman Key Exchange Scheme, Each Party Generates a Public/ Private Key Pair and Distributes the Public Key of the Pair. After Obtaining an Authentic (N.B., this is Critical) Copy of Each other's Public Keys, Alice and Bob can Compute a Shared Secret Offline. The Shared Secret can be Used, for Instance, as the Key for a Symmetric Cipher

- In a public-key encryption system, anyone with a public key can encrypt a message, yielding a ciphertext, but only those who know the corresponding private key can decrypt the ciphertext to obtain the original message. For example, a journalist can publish the public key of an encryption key pair on a web site so that sources can send secret messages to the news organization in ciphertext. Only the journalist who knows the corresponding private key can decrypt the ciphertexts to obtain the sources' messages—an eavesdropper reading email on its way to the journalist cannot decrypt the ciphertexts. However, public-key encryption does not conceal metadata like what computer a source used to send a message, when they sent it, or how long it is. Public-key encryption on its own also does not tell the recipient anything about who sent a message—it just conceals the content of the message.

One important issue is confidence/proof that a particular public key is authentic, i.e. that it is correct and belongs to the person or entity claimed, and has not been tampered with or replaced by some (perhaps malicious) third party. There are several possible approaches, including: A public key infrastructure (PKI), in which one or more third parties - known as certificate authorities - certify ownership of key pairs. TLS relies upon this. This implies that the PKI system (software, hardware, and management) is trust-able by all involved.

A "web of trust" decentralizes authentication by using individual endorsements of links between a user and the public key belonging to that user. PGP uses this approach, in addition to lookup in the domain name system (DNS). The DKIM system for digitally signing emails also uses this approach.

Applications

The most obvious application of a public key encryption system is for encrypting communication to provide confidentiality - a message that a sender encrypts using the recipient's public key, which can be decrypted only by the recipient's paired private key. Most digital services such as financial services, email, and messaging applications utilized daily are secured using public key encryption.

Another application in public key cryptography is the digital signature. Digital signature schemes can be used for sender authentication. Non-repudiation systems use digital signatures to ensure that one party cannot successfully dispute its authorship of a document or communication. Further applications built on this foundation include: digital cash, password-authenticated key agreement, time-stamping services and non-repudiation protocols.

Hybrid Cryptosystems

Because asymmetric key algorithms are nearly always much more computationally intensive than symmetric ones, it is common to use a public/private asymmetric key-exchange algorithm to encrypt and exchange a symmetric key, which is then used by symmetric-key cryptography to transmit data using the now-shared symmetric key for a symmetric key encryption algorithm. PGP, SSH, and the SSL/TLS family of schemes use this procedure; they are thus called hybrid cryptosystems. The initial asymmetric cryptography-based key exchange to share a server-generated symmetric key from the server to client has the advantage of not requiring that a symmetric key be pre-shared manually, such as on printed paper or discs transported by a courier, while providing the higher data throughput of symmetric key cryptography over asymmetric key cryptography for the remainder of the shared connection.

Weaknesses

As with all security-related systems, there are various potential weaknesses in public-key cryptography. Aside from poor choice of an asymmetric key algorithm (there are few that are widely regarded as satisfactory) or too short a key length, the chief security risk is that the private key of a pair becomes known. All security of messages, authentication, etc., will then be lost. Additionally, with the advent of quantum computing, many asymmetric key algorithms are considered vulnerable to attacks, and new quantum-resistant schemes are being developed to overcome the problem.

Algorithms

All public key schemes are in theory susceptible to a "brute-force key search attack". However, such an attack is impractical if the amount of computation needed to succeed – termed the "work factor" by Claude Shannon – is out of reach of all potential attackers. In many cases, the work factor can be increased by simply choosing a longer key. But other algorithms may inherently have much lower work factors, making resistance to a brute-force attack (e.g., from longer keys) irrelevant. Some special and specific algorithms have been developed to aid in attacking some public key encryption algorithms; both RSA and ElGamal encryption have known attacks that are much faster than the brute-force approach. None of these are sufficiently improved to be actually practical, however.

Major weaknesses have been found for several formerly promising asymmetric key algorithms. The "knapsack packing" algorithm was found to be insecure after the development of a new attack. As with all cryptographic functions, public-key implementations may be vulnerable to side-channel attacks that exploit information leakage to simplify the search for a secret key. These are often independent of the algorithm being used. Research is underway to both discover, and to protect against, new attacks.

Alteration of Public Keys

Another potential security vulnerability in using asymmetric keys is the possibility of a "man-in-the-middle" attack, in which the communication of public keys is intercepted by a third party (the "man in the middle") and then modified to provide different public keys instead. Encrypted messages and responses must, in all instances, be intercepted, decrypted, and re-encrypted by the attacker using the correct public keys for the different communication segments so as to avoid suspicion. A communication is said to be insecure where data is transmitted in a manner that allows for interception (also called "sniffing"). These terms refer to reading the sender's private data in its entirety. A communication is particularly unsafe when interceptions can not be prevented or monitored by the sender.

A man-in-the-middle attack can be difficult to implement due to the complexities of modern security protocols. However, the task becomes simpler when a sender is using insecure media such as public networks, the Internet, or wireless communication. In these cases an attacker can compromise the communications infrastructure rather than the data itself. A hypothetical malicious staff member at an Internet service provider (ISP) might find a man-in-the-middle attack relatively straightforward. Capturing the public key would only require searching for the key as it gets sent through the ISP's communications hardware; in properly implemented asymmetric key schemes, this is not a significant risk.

In some advanced man-in-the-middle attacks, one side of the communication will see the original data while the other will receive a malicious variant. Asymmetric man-in-the-middle attacks can prevent users from realizing their connection is compromised. This remains so even when one user's data is known to be compromised because the data appears fine to the other user. This can lead to confusing disagreements between users such as "it must be on your end!" when neither user is at fault. Hence, man-in-the-middle attacks are only fully preventable when the communications infrastructure is physically controlled by one or both parties; such as via a wired route inside the sender's own building. In summation, public keys are easier to alter when the communications hardware used by a sender is controlled by an attacker.

Public Key Infrastructure

One approach to prevent such attacks involves the use of a public key infrastructure (PKI); a set of roles, policies, and procedures needed to create, manage, distribute, use, store and revoke digital certificates and manage public-key encryption. However, this has potential weaknesses. For example, the certificate authority issuing the certificate must be trusted by all participating parties to have properly checked the identity of the key-holder, to have ensured the correctness of the public key when it issues a certificate, to be secure from computer piracy, and to have made arrangements with all participants to check all their certificates before protected communications can begin. Web browsers, for instance, are supplied with a long list of "self-signed identity certificates" from PKI providers – these are used to check the bona fides of the certificate authority and then, in a second step, the certificates of potential communicators. An attacker who could subvert one of those certificate authorities into issuing a certificate for a bogus public key could then mount a "man-in-the-middle" attack as easily as if the certificate scheme were not used at all. An attacker who penetrates an authority's servers and obtains its store of certificates and keys (public and private) would be able to spoof, masquerade, decrypt, and forge transactions without limit, assuming that they were able to place themselves in the communication stream.

Despite its theoretical and potential problems, Public key infrastructure is widely used. Examples include TLS and its predecessor SSL, which are commonly used to provide security for web browser transactions (for example, most websites utilize TLS for HTTPS). Aside from the resistance to attack of a particular key pair, the security of the certification hierarchy must be considered when deploying public key systems. Some certificate authority – usually a purpose-built program running on a server computer – vouches for the identities assigned to specific private keys by producing a digital certificate. Public key digital certificates are typically valid for several years at a time, so the associated private keys must be held securely over that time. When a private key used for certificate creation higher in the PKI server hierarchy is compromised, or accidentally disclosed, then a "man-in-the-middle attack" is possible, making any subordinate certificate wholly insecure.

Unencrypted Metadata

Most of the available public-key encryption software does not conceal metadata in the message header, which might include the identities of the sender and recipient, the sending date, subject field, and the software they use etc. Rather, only the body of the message is concealed and can only be decrypted with the private key of the intended recipient. This means that a third party could construct quite a detailed model of participants in a communication network, along with the subjects being discussed, even if the message body itself is hidden. However, there has been a recent demonstration of messaging with encrypted headers, which obscures the identities of the sender and recipient, and significantly reduces the available metadata to a third party. The concept is based around an open repository containing separately encrypted metadata blocks and encrypted messages. Only the intended recipient is able to decrypt the metadata block, and having done so they can identify and download their messages and decrypt them. Such a messaging system is at present in an experimental phase and not yet deployed. Scaling this method would reveal to the third party only the inbox server being used by the recipient and the timestamp of sending and receiving. The server could be shared by thousands of users, making social network modelling much more challenging.

2.1.2 Symmetric-Key

Symmetric-key algorithms are algorithms for cryptography that use the same cryptographic keys for both the encryption of plaintext and the decryption of ciphertext. The keys may be identical, or there may be a simple transformation to go between the two keys. The keys, in practice, represent a shared secret between two or more parties that can be used to maintain a private information link. The requirement that both parties have access to the secret key is one of the main drawbacks of symmetric-key encryption, in comparison to public-key encryption (also known as asymmetric-key encryption). However, symmetric-key encryption algorithms are usually better for bulk encryption. With exception of the one-time pad they have a smaller key size, which means less storage space and faster transmission. Due to this, asymmetric-key encryption is often used to exchange the secret key for symmetric-key encryption.

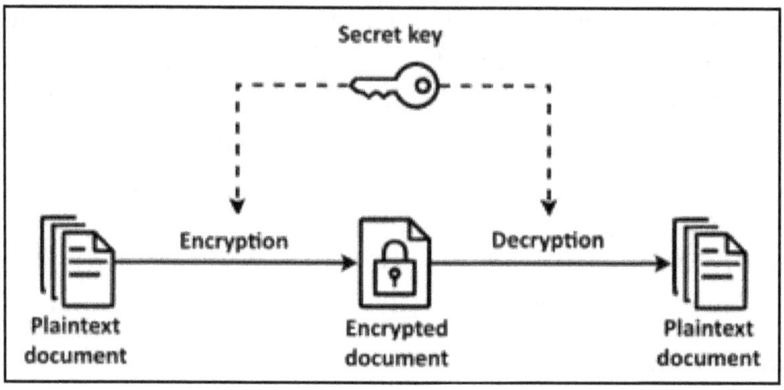

Fig: Symmetric-Key Encryption: The Same Key is used for Both Encryption and Decryption

Types

- Symmetric-key encryption can use either stream ciphers or block ciphers.
- Stream ciphers encrypt the digits (typically bytes), or letters (in substitution ciphers) of a message one at a time. An example is ChaCha20. Substitution ciphers are well-known ciphers, but can be easily decrypted using a frequency table.
- Block ciphers take a number of bits and encrypt them in a single unit, padding the plaintext to achieve a multiple of the block size. The Advanced Encryption Standard (AES) algorithm, approved by NIST in December 2001, uses 128-bit blocks.

Implementations

Examples of popular symmetric-key algorithms include:

- Twofish,
- Serpent,

- AES (Rijndael),
- Camellia,
- Salsa20,
- ChaCha20,
- Blowfish,
- CAST5,
- Kuznyechik,
- RC4,
- DES,
- 3DES,
- Skipjack,
- Safer, and
- IDEA.

Use as a Cryptographic Primitive

Symmetric ciphers are commonly used to achieve other cryptographic primitives than just encryption. Encrypting a message does not guarantee that it will remain unchanged while encrypted. Hence, often a message authentication code is added to a ciphertext to ensure that changes to the ciphertext will be noted by the receiver. Message authentication codes can be constructed from an AEAD cipher (e.g. AES-GCM). However, symmetric ciphers cannot be used for non-repudiation purposes except by involving additional parties. Another application is to build hash functions from block ciphers.

Construction of Symmetric Ciphers

Many modern block ciphers are based on a construction proposed by Horst Feistel. Feistel's construction makes it possible to build invertible functions from other functions that are themselves not invertible.

Security of Symmetric Ciphers

Symmetric ciphers have historically been susceptible to known-plaintext attacks, chosen-plaintext attacks, differential cryptanalysis and linear cryptanalysis. Careful construction of the functions for each round can greatly reduce the chances of a successful attack. It is also possible to increase the key length or the rounds in the encryption process to better protect against attack. This, however, tends to increase the processing power and decrease the speed at which the process runs due to the amount of operations the system needs to do.

Most modern symmetric-key algorithms appear to be resistant to the threat of post-quantum cryptography. Quantum computers would exponentially increase the speed at which these ciphers can be decoded; notably, Grover's algorithm would take the square-root of the time traditionally required for a brute-force attack, although these vulnerabilities can be compensated for by doubling key length. For example, a 128 bit AES cipher would not be secure against such an attack as it would reduce the time required to test all possible iterations from over 10 quintillion years to about six months. By contrast, it would still take a quantum computer the same amount of time to decode a 256 bit AES cipher as it would a conventional computer to decode a 128 bit AES cipher. For this reason, AES-256 is believed to be "quantum resistant".

Key Establishment

Symmetric-key algorithms require both the sender and the recipient of a message to have the same secret key. All early cryptographic systems required either the sender or the recipient to somehow receive a copy of that secret key over a physically secure channel. Nearly all modern cryptographic systems still use symmetric-key algorithms internally to encrypt the bulk of the messages, but they eliminate the need for a physically secure channel by using Diffie–Hellman key exchange or some other public-key protocol to securely come to agreement on a fresh new secret key for each session/conversation (forward secrecy).

Key Generation

When used with asymmetric ciphers for key transfer, pseudorandom key generators are nearly always used to generate the symmetric cipher session keys. However, lack of randomness in those generators or in their initialization vectors is disastrous and has led to cryptanalytic breaks in the past. Therefore, it is essential that an implementation use a source of high entropy for its initialization.

Reciprocal Cipher

A reciprocal cipher is a cipher where, just as one enters the plaintext into the cryptography system to get the ciphertext, one could enter the ciphertext into the same place in the system to get the plaintext. A reciprocal cipher is also sometimes referred as self-reciprocal cipher. Practically all mechanical cipher machines implement a reciprocal cipher, a mathematical involution on each typed-in letter. Instead of designing two kinds of machines, one for encrypting and one for decrypting, all the machines can be identical and can be set up (keyed) the same way. Examples of reciprocal ciphers include:

- Atbash.
- Beaufort cipher.

- Enigma machine.
- Marie Antoinette and Axel von Fersen communicated with a self-reciprocal cipher.
- the Porta polyalphabetic cipher is self-reciprocal.
- Purple cipher.
- RC4.
- ROT13.
- XOR cipher.
- Vatsyayana cipher.

The majority of all modern ciphers can be classified as either a stream cipher, most of which use a reciprocal XOR cipher combiner, or a block cipher, most of which use a Feistel cipher or Lai–Massey scheme with a reciprocal transformation in each round.

2.1.3 Hash Functions

A hash function is any function that can be used to map data of arbitrary size to fixed-size values, though there are some hash functions that support variable-length output. The values returned by a hash function are called hash values, hash codes, (hash/message) digests, or simply hashes. The values are usually used to index a fixed-size table called a hash table. Use of a hash function to index a hash table is called hashing or scatter-storage addressing.

Hash functions and their associated hash tables are used in data storage and retrieval applications to access data in a small and nearly constant time per retrieval. They require an amount of storage space only fractionally greater than the total space required for the data or records themselves. Hashing is a computationally- and storage-space-efficient form of data access that avoids the non-constant access time of ordered and unordered lists and structured trees, and the often-exponential storage requirements of direct access of state spaces of large or variable-length keys.

Use of hash functions relies on statistical properties of key and function interaction: worst-case behavior is intolerably bad but rare, and average-case behavior can be nearly optimal (minimal collision). Hash functions are related to (and often confused with) checksums, check digits, fingerprints, lossy compression, randomization functions, error-correcting codes, and ciphers.

Although the concepts overlap to some extent, each one has its own uses and requirements and is designed and optimized differently. The hash function differs from these concepts mainly in terms of data integrity. Hash tables may use non-cryptographic hash functions, while cryptographic hash functions are used in cybersecurity to secure sensitive data such as passwords. In a hash table, a hash function takes a key as an input, which is associated with a datum or record and used to identify it to the data storage and retrieval application. The keys may be fixed-length, like an integer, or variable-length, like a name. In some cases, the key is the datum itself. The output is a hash code used to index a hash table holding the data or records, or pointers to them. A hash function may be considered to perform three functions:

- Convert variable-length keys into fixed-length (usually machine-word-length or less) values, by folding them by words or other units using a parity-preserving operator like ADD or XOR,
- Scramble the bits of the key so that the resulting values are uniformly distributed over the keyspace, and
- Map the key values into ones less than or equal to the size of the table.

A good hash function satisfies two basic properties: it should be very fast to compute, and it should minimize duplication of output values (collisions). Hash functions rely on generating favorable probability distributions for their effectiveness, reducing access time to nearly constant. High table loading factors, pathological key sets, and poorly designed hash functions can result in access times approaching linear in the number of items in the table. Hash functions can be designed to give the best worst-case performance, good performance under high table loading factors, and in special cases, perfect (collisionless) mapping of keys into hash codes. Implementation is based on parity-preserving bit operations (XOR and ADD), multiply, or divide. A necessary adjunct to the hash function is a collision-resolution method that employs an auxiliary data structure like linked lists, or systematic probing of the table to find an empty slot.

Hash Tables

Hash functions are used in conjunction with hash tables to store and retrieve data items or data records. The hash function translates the key associated with each datum or record into a hash code, which is used to index the hash table. When an item is to be added to the table, the hash code may index an empty slot (also called a bucket), in which case the item is added to the table there. If the hash code indexes a full slot, then some kind of collision resolution is required: the new item may be omitted (not added to the table), or replace the old item, or be added to the table in some other location by a specified procedure.

That procedure depends on the structure of the hash table. In chained hashing, each slot is the head of a linked list or chain, and items that collide at the slot are added to the chain. Chains may be kept in random order and searched linearly, or in serial order, or as a self-ordering list by frequency to speed up access. In open address hashing, the table is probed starting from the occupied slot in a specified manner, usually by linear probing, quadratic probing, or double hashing until an open slot is located or the entire table is probed (overflow). Searching for the item follows the same procedure until the item is located, an open slot is found, or the entire table has been searched (item not in table).

Specialized Uses

Hash functions are also used to build caches for large data sets stored in slow media. A cache is generally simpler than a hashed search table, since any collision can be resolved by discarding or writing back the older of the two colliding items. Hash functions are an essential ingredient of the Bloom filter, a space-efficient probabilistic data structure that is used to test whether an element is a member of a set.

A special case of hashing is known as geometric hashing or the grid method. In these applications, the set of all inputs is some sort of metric space, and the hashing function can be interpreted as a partition of that space into a grid of cells. The table is often an array with two or more indices (called a grid file, grid index, bucket grid, and similar names), and the hash function returns an index tuple. This principle is widely used in computer graphics, computational geometry, and many other disciplines, to solve many proximity problems in the plane or in three-dimensional space, such as finding closest pairs in a set of points, similar shapes in a list of shapes, similar images in an image database, and so on. Hash tables are also used to implement associative arrays and dynamic sets.

Properties

Uniformity

A good hash function should map the expected inputs as evenly as possible over its output range. That is, every hash value in the output range should be generated with roughly the same probability. The reason for this last requirement is that the cost of hashing-based methods goes up sharply as the number of collisions—pairs of inputs that are mapped to the same hash value—increases. If some hash values are more likely to occur than others, then a larger fraction of the lookup operations will have to search through a larger set of colliding table entries. This criterion only requires the value to be uniformly distributed, not random in any sense. A good randomizing function is (barring computational efficiency concerns) generally a good choice as a hash function, but the converse need not be true.

Hash tables often contain only a small subset of the valid inputs. For instance, a club membership list may contain only a hundred or so member names, out of the very large set of all possible names. In these cases, the uniformity criterion should hold for almost all typical subsets of entries that may be found in the table, not just for the global set of all possible entries. In other words, if a typical set of m records is hashed to n table slots, then the probability of a bucket receiving many more than m/n records should be vanishingly small. In particular, if m < n, then very few buckets should have more than one or two records. A small number of collisions is virtually inevitable, even if n is much larger than m.

In special cases when the keys are known in advance and the key set is static, a hash function can be found that achieves absolute (or collisionless) uniformity. Such a hash function is said to be perfect. There is no algorithmic way of constructing such a function—searching for one is a factorial function of the number of keys to be mapped versus the number of table slots that they are mapped into. Finding a perfect hash function over more than a very small set of keys is usually computationally infeasible; the resulting function is likely to be more computationally complex than a standard hash function and provides only a marginal advantage over a function with good statistical properties that yields a minimum number of collisions.

Testing and Measurement

When testing a hash function, the uniformity of the distribution of hash values can be evaluated by the chi-squared test. This test is a goodness-of-fit measure: it is the actual distribution of items in buckets versus the expected (or uniform) distribution of items. The formula is

$$\frac{\sum_{j=0}^{m-1}(b_j)(b_j+1)/2}{(n/2m)(n+2m-1)},$$

where n is the number of keys, m is the number of buckets, and b_j is the number of items in bucket j. A ratio within one confidence interval (such as 0.95 to 1.05) is indicative that the hash function evaluated has an expected uniform distribution. Hash functions can have some technical properties that make it more likely that they will have a uniform distribution when applied. One is the strict avalanche criterion: whenever a single input bit is complemented, each of the output bits changes with a 50% probability. The reason for this property is that selected subsets of the keyspace may have low variability.

For the output to be uniformly distributed, a low amount of variability, even one bit, should translate into a high amount of variability (i.e. distribution over the tablespace) in the output. Each bit should change with a probability of 50% because, if some bits are reluctant to change, then the keys become clustered around those values. If the bits want to change too readily, then the mapping is approaching a fixed XOR function of a single bit. Standard tests for this property have been described in the literature. The relevance of the criterion to a multiplicative hash function is assessed here.

Efficiency

In data storage and retrieval applications, the use of a hash function is a trade-off between search time and data storage space. If search time were unbounded, then a very compact unordered linear list would be the best medium; if storage space were unbounded, then a randomly accessible structure indexable by the key-value would be very large and very sparse, but very fast. A hash function takes a finite amount of time to map a potentially large keyspace to a feasible amount of storage space searchable in a bounded amount of time regardless of the number of keys. In most applications, the hash function should be computable with minimum latency and secondarily in a minimum number of instructions.

Computational complexity varies with the number of instructions required and latency of individual instructions, with the simplest being the bitwise methods (folding), followed by the multiplicative methods, and the most complex (slowest) are the division-based methods. Because collisions should be infrequent, and cause a marginal delay but are otherwise harmless, it is usually preferable to choose a faster hash function over one that needs more computation but saves a few collisions.

Division-based implementations can be of particular concern because a division requires multiple cycles on nearly all processor microarchitectures. Division (modulo) by a constant can be inverted to become a multiplication by the word-size multiplicative-inverse of that constant. This can be done by the programmer, or by the compiler. Division can also be reduced directly into a series of shift-subtracts and shift-adds, though minimizing the number of such operations required is a daunting problem; the number of machine-language instructions resulting may be more than a dozen and swamp the pipeline. If the microarchitecture has hardware multiply functional units, then the multiply-by-inverse is likely a better approach.

We can allow the table size n to not be a power of 2 and still not have to perform any remainder or division operation, as these computations are sometimes costly. For example, let n be significantly less than 2^b. Consider a pseudorandom number generator function P(key) that is uniform on the interval $[0, 2^b - 1]$. A hash function uniform on the interval $[0, n - 1]$ is n P(key) / 2^b. We can replace the division by a (possibly faster) right bit shift: n P(key) >> b.

If keys are being hashed repeatedly, and the hash function is costly, then computing time can be saved by precomputing the hash codes and storing them with the keys. Matching hash codes almost certainly means that the keys are identical. This technique is used for the transposition table in game-playing programs, which stores a 64-bit hashed representation of the board position.

Universality

A universal hashing scheme is a randomized algorithm that selects a hash function h among a family of such functions, in such a way that the probability of a collision of any two distinct keys is 1/m, where m is the number of distinct hash values desired—independently of the two keys. Universal hashing ensures (in a probabilistic sense) that the hash function application will behave as well as if it were using a random function, for any distribution of the input data. It will, however, have more collisions than perfect hashing and may require more operations than a special-purpose hash function.

Applicability

A hash function that allows only certain table sizes or strings only up to a certain length, or cannot accept a seed (i.e. allow double hashing) is less useful than one that does. A hash function is applicable in a variety of situations. Particularly within cryptography, notable applications include:

- **Integrity Checking:** Identical hash values for different files imply equality, providing a reliable means to detect file modifications.
- **Key Derivation:** Minor input changes result in a random-looking output alteration, known as the diffusion property. Thus, hash functions are valuable for key derivation functions.
- **Message Authentication Codes (MACs):** Through the integration of a confidential key with the input data, hash functions can generate MACs ensuring the genuineness of the data, such as in HMACs.
- **Password Storage:** The password's hash value does not expose any password details, emphasizing the importance of securely storing hashed passwords on the server.
- **Signatures:** Message hashes are signed rather than the whole message.

Deterministic

A hash procedure must be deterministic—for a given input value, it must always generate the same hash value. In other words, it must be a function of the data to be hashed, in the mathematical sense of the term. This requirement excludes hash functions that depend on external variable parameters, such as pseudo-random number generators or the time of day. It also excludes functions that depend on the memory address of the object being hashed, because the address may change during execution (as may happen on systems that use certain methods of garbage collection), although sometimes rehashing of the item is possible.

The determinism is in the context of the reuse of the function. For example, Python adds the feature that hash functions make use of a randomized seed that is generated once when the Python process starts in addition to the input to be hashed. The Python hash (SipHash) is still a valid hash function when used within a single run, but if the values are persisted (for example, written to disk), they can no longer be treated as valid hash values, since in the next run the random value might differ.

Defined Range

It is often desirable that the output of a hash function have fixed size. If, for example, the output is constrained to 32-bit integer values, then the hash values can be used to index into an array. Such hashing is commonly used to accelerate data searches. Producing fixed-length output from variable-length input can be accomplished by breaking the input data into chunks of specific size. Hash functions used for data searches use some arithmetic expression that iteratively processes chunks of the input (such as the characters in a string) to produce the hash value.

Variable Range

In many applications, the range of hash values may be different for each run of the program or may change along the same run (for instance, when a hash table needs to be expanded). In those situations, one needs a hash function which takes two parameters—the input data z, and the number n of allowed hash values. A common solution is to compute a fixed hash function with a very large range (say, 0 to $2^{32} - 1$), divide the result by n, and use the division's remainder. If n is itself a power of 2, this can be done by bit masking and bit shifting. When this approach is used, the hash function must be chosen so that the result has fairly uniform distribution between 0 and n - 1, for any value of n that may occur in the application. Depending on the function, the remainder may be uniform only for certain values of n, e.g. odd or prime numbers.

Variable Range with Minimal Movement (Dynamic Hash Function)

When the hash function is used to store values in a hash table that outlives the run of the program, and the hash table needs to be expanded or shrunk, the hash table is referred to as a dynamic hash table. A hash function that will relocate the minimum number of records when the table is resized is desirable. What is needed is a hash function $H(z,n)$ (where z is the key being hashed and n is the number of allowed hash values) such that $H(z,n + 1) = H(z,n)$ with probability close to $n/(n + 1)$.

Linear hashing and spiral hashing are examples of dynamic hash functions that execute in constant time but relax the property of uniformity to achieve the minimal movement property. Extendible hashing uses a dynamic hash function that requires space proportional to n to compute the hash function, and it becomes a function of the previous keys that have been inserted. Several algorithms that preserve the uniformity property but require time proportional to n to compute the value of H(z,n) have been invented. A hash function with minimal movement is especially useful in distributed hash tables.

Data Normalization

In some applications, the input data may contain features that are irrelevant for comparison purposes. For example, when looking up a personal name, it may be desirable to ignore the distinction between upper and lower case letters. For such data, one must use a hash function that is compatible with the data equivalence criterion being used: that is, any two inputs that are considered equivalent must yield the same hash value. This can be accomplished by normalizing the input before hashing it, as by upper-casing all letters.

Hashing Integer Data Types

There are several common algorithms for hashing integers. The method giving the best distribution is data-dependent. One of the simplest and most common methods in practice is the modulo division method.

Identity Hash Function

If the data to be hashed is small enough, then one can use the data itself (reinterpreted as an integer) as the hashed value. The cost of computing this identity hash function is effectively zero. This hash function is perfect, as it maps each input to a distinct hash value. The meaning of "small enough" depends on the size of the type that is used as the hashed value. For example, in Java, the hash code is a 32-bit integer. Thus the 32-bit integer Integer and 32-bit floating-point Float objects can simply use the value directly, whereas the 64-bit integer Long and 64-bit floating-point Double cannot. Other types of data can also use this hashing scheme. For example, when mapping character strings between upper and lower case, one can use the binary encoding of each character, interpreted as an integer, to index a table that gives the alternative form of that character ("A" for "a", "8" for "8", etc.). If each character is stored in 8 bits (as in extended ASCII or ISO Latin 1), the table has only $2^8 = 256$ entries; in the case of Unicode characters, the table would have $17 \times 2^{16} = 1114112$ entries. The same technique can be used to map two-letter country codes like "us" or "za" to country names ($26^2 = 676$ table entries), 5-digit ZIP codes like 13083 to city names (100000 entries), etc. Invalid data values (such as the country code "xx" or the ZIP code 00000) may be left undefined in the table or mapped to some appropriate "null" value.

Trivial Hash Function

If the keys are uniformly or sufficiently uniformly distributed over the key space, so that the key values are essentially random, then they may be considered to be already "hashed". In this case, any number of any bits in the key may be extracted and collated as an index into the hash table. For example, a simple hash function might mask off the m least significant bits and use the result as an index into a hash table of size 2^m.

Mid-Squares

A mid-squares hash code is produced by squaring the input and extracting an appropriate number of middle digits or bits. For example, if the input is 123456789 and the hash table size 10000, then squaring the key produces 15241578750190521, so the hash code is taken as the middle 4 digits of the 17-digit number (ignoring the high digit) 8750. The mid-squares method produces a reasonable hash code if there is not a lot of leading or trailing zeros in the key. This is a variant of multiplicative hashing, but not as good because an arbitrary key is not a good multiplier.

Division Hashing

A standard technique is to use a modulo function on the key, by selecting a divisor M which is a prime number close to the table size, so $h(K) \equiv K$ (mod M). The table size is usually a power of 2. This gives a distribution from {0, M - 1}. This gives good results over a large number of key sets. A significant drawback of division hashing is that division requires multiple cycles on most modern architectures (including x86) and can be 10 times slower than multiplication. A second drawback is that it will not break up clustered keys. For example, the keys 123000, 456000, 789000, etc. modulo 1000 all map to the same address. This technique works well in practice because many key sets are sufficiently random already, and the probability that a key set will be cyclical by a large prime number is small.

Algebraic Coding

Algebraic coding is a variant of the division method of hashing which uses division by a polynomial modulo 2 instead of an integer to map n bits to m bits. In this approach, $M = 2^m$, and we postulate an mth-degree polynomial $Z(x) = x^m + \zeta_{m-1}x^{m-1} + \cdots + \zeta_0$. A key $K = (k_{n-1}...k_1k_0)_2$ can be regarded as the polynomial $K(x) = k_{n-1}x^{n-1} + \cdots + k_1x + k_0$. The remainder using polynomial arithmetic modulo 2 is K(x) mod $Z(x) = h_{m-1}x^{m-1} + \cdots h_1x + h_0$. Then $h(K) = (h_{m-1}...h_1h_0)_2$. If Z(x) is constructed to have t or fewer non-zero coefficients, then keys which share fewer than t bits are guaranteed to not collide. Z is a function of k, t, and n (the last of which is a divisor of $2^k - 1$) and is constructed from the finite field $GF(2^k)$. Knuth gives an example: taking (n,m,t) = (15,10,7) yields $Z(x) = x^{10} + x^8 + x^5 + x^4 + x^2 + x + 1$. The derivation is as follows:

Let S be the smallest set of integers such that $\{1,2,...,t\} \subseteq S$ and $(2j \bmod n) \in S \ \forall j \in S$.

Define $P(x) = \prod_{j \in S}(x - \alpha^j)$ where $\alpha \in^n GF(2^k)$ and where the coefficients of P(x) are computed in this field. Then the degree of $P(x) = |S|$. Since α^{2j} is a root of P(x) whenever α^j is a root, it follows that the coefficients p^i of P(x) satisfy p^2

$_{i = pi}$, so they are all 0 or 1. If $R(x) = r_{n-1}x^{n-1} + \cdots + r_1x + r_0$ is any nonzero polynomial modulo 2 with at most t nonzero coefficients, then R(x) is not a multiple of P(x) modulo 2. If follows that the corresponding hash function will map keys with fewer than t bits in common to unique indices. The usual outcome is that either n will get large, or t will get large, or both, for the scheme to be computationally feasible. Therefore, it is more suited to hardware or microcode implementation.

Multiplicative Hashing

Standard multiplicative hashing uses the formula $h_a(K) = \lfloor (aK \bmod W) / (W/M) \rfloor$, which produces a hash value in {0, ..., M - 1}. The value a is an appropriately chosen value that should be relatively prime to W; it should be large, and its binary representation a random mix of 1s and 0s. An important practical special case occurs when $W = 2^w$ and $M = 2^m$ are powers of 2 and w is the machine word size. In this case, this formula becomes $h_a(K) = \lfloor (aK \bmod 2^w) / 2^{w-m} \rfloor$. This is special because arithmetic modulo 2^w is done by default in low-level programming languages and integer division by a power of 2 is simply a right-shift, so, in C, for example, this function becomes

```
unsigned hash(unsigned K) {
   return (a * K) >> (w - m);
}
```

and for fixed m and w this translates into a single integer multiplication and right-shift, making it one of the fastest hash functions to compute. Multiplicative hashing is susceptible to a "common mistake" that leads to poor diffusion—higher-value input bits do not affect lower-value output bits. A transmutation on the input which shifts the span of retained top bits down and XORs or ADDs them to the key before the multiplication step corrects for this. The resulting function looks like:

```
unsigned hash(unsigned K) {
   K ^= K >> (w - m);
   return (a * K) >> (w - m);
}
```

Fibonacci Hashing

Fibonacci hashing is a form of multiplicative hashing in which the multiplier is $2^w / \phi$, where w is the machine word length and ϕ (phi) is the golden ratio (approximately 1.618). A property of this multiplier is that it uniformly distributes over the table space, blocks of consecutive keys with respect to any block of bits in the key. Consecutive keys within the high bits or low bits of the key (or some other field) are relatively common. The multipliers for various word lengths are:

- 16: $a = 9E37_{16} = 40503_{10}$

- 32: a = $9E3779B9_{16} = 2654435769_{10}$
- 48: a = $9E3779B97F4B_{16} = 173961102589771_{10}$
- 64: a = $9E3779B97F4A7C15_{16} = 11400714819323198485_{10}$

The multiplier should be odd, so the least significant bit of the output is invertible modulo 2^w. The last two values given are rounded (up and down, respectively) by more than 1/2 of a least-significant bit to achieve this.

Zobrist Hashing

Tabulation hashing, more generally known as Zobrist hashing after Albert Zobrist, is a method for constructing universal families of hash functions by combining table lookup with XOR operations. This algorithm has proven to be very fast and of high quality for hashing purposes (especially hashing of integer-number keys).

Zobrist hashing was originally introduced as a means of compactly representing chess positions in computer game-playing programs. A unique random number was assigned to represent each type of piece (six each for black and white) on each space of the board. Thus a table of 64×12 such numbers is initialized at the start of the program. The random numbers could be any length, but 64 bits was natural due to the 64 squares on the board. A position was transcribed by cycling through the pieces in a position, indexing the corresponding random numbers (vacant spaces were not included in the calculation) and XORing them together (the starting value could be 0 (the identity value for XOR) or a random seed). The resulting value was reduced by modulo, folding, or some other operation to produce a hash table index. The original Zobrist hash was stored in the table as the representation of the position.

Later, the method was extended to hashing integers by representing each byte in each of 4 possible positions in the word by a unique 32-bit random number. Thus, a table of 2^8×4 random numbers is constructed. A -32bit hashed integer is transcribed by successively indexing the table with the value of each byte of the plain text integer and XORing the loaded values together (again, the starting value can be the identity value or a random seed). The natural extension to -64bit integers is by use of a table of 2^8×-64 8bit random numbers.

This kind of function has some nice theoretical properties, one of which is called 3-tuple independence, meaning that every 3-tuple of keys is equally likely to be mapped to any 3-tuple of hash values.

Customized Hash Function

A hash function can be designed to exploit existing entropy in the keys. If the keys have leading or trailing zeros, or particular fields that are unused, always zero or some other constant, or generally vary little, then masking out only the volatile bits and hashing on those will provide a better and possibly faster hash function. Selected divisors or multipliers in the division and multiplicative schemes may make more uniform hash functions if the keys are cyclic or have other redundancies.

Hashing Variable-Length Data

When the data values are long (or variable-length) character strings—such as personal names, web page addresses, or mail messages—their distribution is usually very uneven, with complicated dependencies. For example, text in any natural language has highly non-uniform distributions of characters, and character pairs, characteristic of the language. For such data, it is prudent to use a hash function that depends on all characters of the string—and depends on each character in a different way.

Middle and Ends

Simplistic hash functions may add the first and last n characters of a string along with the length, or form a word-size hash from the middle 4 characters of a string. This saves iterating over the (potentially long) string, but hash functions that do not hash on all characters of a string can readily become linear due to redundancies, clustering, or other pathologies in the key set. Such strategies may be effective as a custom hash function if the structure of the keys is such that either the middle, ends, or other fields are zero or some other invariant constant that does not differentiate the keys; then the invariant parts of the keys can be ignored.

Character Folding

The paradigmatic example of folding by characters is to add up the integer values of all the characters in the string. A better idea is to multiply the hash total by a constant, typically a sizable prime number, before adding in the next character, ignoring overflow. Using exclusive-or instead of addition is also a plausible alternative. The final operation would be a modulo, mask, or other function to reduce the word value to an index the size of the table. The weakness of this procedure is that information may cluster in the upper or lower bits of the bytes; this clustering will remain in the hashed result and cause more collisions than a proper randomizing hash. ASCII byte codes, for example, have an upper bit of 0, and printable strings do not use the last byte code or most of the first 32 byte codes, so the information, which uses the remaining byte codes, is clustered in the remaining bits in an unobvious manner.

The classic approach, dubbed the PJW hash based on the work of Peter J. Weinberger at Bell Labs in the 1970s, was originally designed for hashing identifiers into compiler symbol tables as given in the "Dragon Book". This hash function offsets the bytes 4 bits before adding them together. When the quantity wraps, the high 4 bits are shifted out and if non-zero, xored back into the low byte of the cumulative quantity. The result is a word-size hash code to which a modulo or other reducing operation can be applied to produce the final hash index. Today, especially with the advent of 64-bit word sizes, much more efficient variable-length string hashing by word chunks is available.

Word Length Folding

Modern microprocessors will allow for much faster processing if 8-bit character strings are not hashed by processing one character at a time, but by interpreting the string as an array of 32-bit or 64-bit integers and hashing/accumulating these "wide word" integer values by means of arithmetic operations (e.g. multiplication by constant and bit-shifting). The final word, which may have unoccupied byte positions, is filled with zeros or a specified randomizing value before being folded into the hash. The accumulated hash code is reduced by a final modulo or other operation to yield an index into the table.

Radix Conversion Hashing

Analogous to the way an ASCII or EBCDIC character string representing a decimal number is converted to a numeric quantity for computing, a variable-length string can be converted as $x_{k-1}a^{k-1} + x_{k-2}a^{k-2} + \cdots + x_1a + x_0$. This is simply a polynomial in a radix $a > 1$ that takes the components $(x_0,x_1,\ldots,x_{k-1})$ as the characters of the input string of length k.

It can be used directly as the hash code, or a hash function applied to it to map the potentially large value to the hash table size. The value of a is usually a prime number large enough to hold the number of different characters in the character set of potential keys. Radix conversion hashing of strings minimizes the number of collisions. Available data sizes may restrict the maximum length of string that can be hashed with this method. For example, a 128-bit word will hash only a 26-character alphabetic string (ignoring case) with a radix of 29; a printable ASCII string is limited to 9 characters using radix 97 and a 64-bit word. However, alphabetic keys are usually of modest length, because keys must be stored in the hash table. Numeric character strings are usually not a problem; 64 bits can count up to 10^{19}, or 19 decimal digits with radix 10.

Rolling Hash

In some applications, such as substring search, one can compute a hash function h for every k-character substring of a given n-character string by advancing a window of width k characters along the string, where k is a fixed integer, and $n > k$. The straightforward solution, which is to extract such a substring at every character position in the text and compute h separately, requires a number of operations proportional to $k \cdot n$. However, with the proper choice of h, one can use the technique of rolling hash to compute all those hashes with an effort proportional to $mk + n$

where m is the number of occurrences of the substring.

The most familiar algorithm of this type is Rabin-Karp with best and average case performance O(n+mk) and worst case O(n·k) (in all fairness, the worst case here is gravely pathological: both the text string and substring are composed of a repeated single character, such as t="AAAAAAAAAAA", and s="AAA"). The hash function used for the algorithm is usually the Rabin fingerprint, designed to avoid collisions in 8-bit character strings, but other suitable hash functions are also used.

Fuzzy Hash

Fuzzy hashing, also known as similarity hashing, is a technique for detecting data that is similar, but not exactly the same, as other data. This is in contrast to cryptographic hash functions, which are designed to have significantly different hashes for even minor differences. Fuzzy hashing has been used to identify malware and has potential for other applications, like data loss prevention and detecting multiple versions of code.

Perceptual Hash

Perceptual hashing is the use of a fingerprinting algorithm that produces a snippet, hash, or fingerprint of various forms of multimedia. A perceptual hash is a type of locality-sensitive hash, which is analogous if features of the multimedia are similar. This is in contrast to cryptographic hashing, which relies on the avalanche effect of a small change in input value creating a drastic change in output value. Perceptual hash functions are widely used in finding cases of online copyright infringement as well as in digital forensics because of the ability to have a correlation between hashes so similar data can be found (for instance with a differing watermark).

Analysis

Worst case results for a hash function can be assessed two ways: theoretical and practical. The theoretical worst case is the probability that all keys map to a single slot. The practical worst case is the expected longest probe sequence (hash function + collision resolution method). This analysis considers uniform hashing, that is, any key will map to any particular slot with probability 1/m, a characteristic of universal hash functions. While Knuth worries about adversarial attack on real time systems, Gonnet has shown that the probability of such a case is "ridiculously small". His representation was that the probability of k of n keys mapping to a single slot is $\alpha^k / (e^\alpha k!)$, where α is the load factor, n/m.

2.2 CRYPTOGRAPHIC ASSUMPTIONS AND THEIR BREAKABILITY

Cryptographic assumptions are foundational beliefs upon which the security of modern cryptographic systems is built. These assumptions are rooted in computational hardness—that is, certain mathematical problems are assumed to be too difficult for any adversary to solve within a reasonable time frame using current computational resources. Cryptographic protocols such as encryption schemes, digital signatures, and key exchanges rely on these assumptions to guarantee the confidentiality, integrity, and authenticity of digital data. Cryptographic assumptions are essential to the security of digital systems, but they are not infallible. They rest on the belief that certain problems are too complex to solve with current technology. However, history has repeatedly shown that assumptions once deemed secure can be broken by advances in computation, mathematics, or technology.

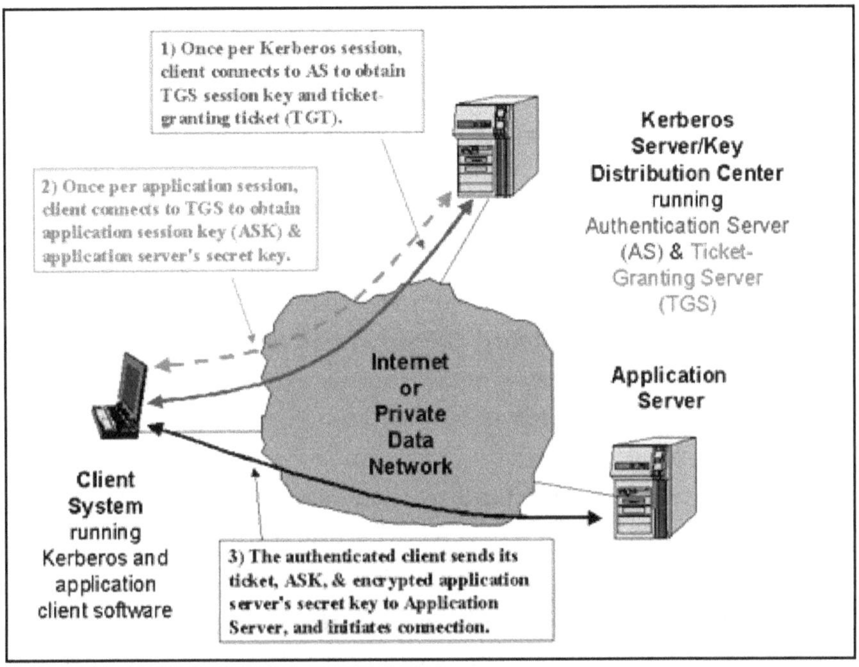

Fig: Cryptographic Assumptions

Quantum computing, improved algorithms, and practical attack techniques continue to challenge these foundational beliefs. Therefore, cryptographers must remain vigilant, adaptive, and collaborative to build systems that are resilient to evolving threats. The future of digital security depends not just on choosing the right assumptions, but on anticipating when and how they might fail.

- **Common Cryptographic Assumptions:** Several widely used cryptographic systems rely on specific hard problems. RSA encryption is based on the assumption that factoring the product of two large prime numbers is computationally infeasible. Similarly, the security of Elliptic Curve Cryptography (ECC) is grounded in the difficulty of solving the elliptic curve discrete logarithm problem. The Diffie-Hellman key exchange assumes the hardness of computing discrete logarithms in a finite field. In each of these cases, the underlying problem is easy to compute in one direction but practically impossible to reverse without special knowledge such as a private key. These one-way functions make secure communication possible, even over untrusted networks.

- **Symmetric-Key Cryptography and Complexity:** In symmetric-key cryptography, the assumptions are somewhat different. The strength of ciphers such as AES is not derived from a specific mathematical problem but from the absence of any known efficient attack better than brute force. The cryptographic community assumes that AES, for example, is secure because no shortcut algorithm has been discovered that could break it within a practical timeframe. This assumption is bolstered by years of cryptanalysis, but it remains conditional—if a better algorithm or implementation flaw is discovered, the assumption may no longer hold.
- **Evolution of Cryptographic Vulnerabilities:** Over time, cryptographic assumptions that once seemed secure have been broken due to advancements in hardware and algorithms. A historical example is the Data Encryption Standard (DES), which was considered secure in the 1970s with a 56-bit key length. However, by the 1990s, the rise of faster computing allowed DES to be broken by brute-force attacks. Similarly, the MD5 and SHA-1 hash functions, once trusted, were eventually broken due to improved understanding of collision attacks. These developments demonstrate that the perceived hardness of cryptographic problems can change, undermining previously held assumptions.
- **Quantum Computing and Its Threat:** Quantum computing presents one of the most serious threats to current cryptographic assumptions. Algorithms like Shor's algorithm, if run on a sufficiently powerful quantum computer, could solve the integer factorization and discrete logarithm problems in polynomial time. This would render RSA, DSA, and ECC-based systems insecure. Although quantum computers capable of executing such algorithms at scale do not yet exist, their potential has prompted the cryptographic community to develop quantum-resistant algorithms. Post-quantum cryptography relies on newer assumptions such as the hardness of lattice problems, multivariate polynomials, and code-based cryptography. While these new assumptions are believed to be secure against both classical and quantum attacks, they are not immune to future discoveries.
- **Mathematical Advances and Cryptanalysis:** Another source of risk to cryptographic assumptions comes from mathematical advances. As researchers develop new techniques in number theory, algebra,

and computational complexity, previously difficult problems may become tractable. For example, improvements in integer factorization algorithms like the Number Field Sieve have reduced the time required to break RSA keys of certain lengths. If a breakthrough occurs that allows one of these problems to be solved efficiently, the corresponding cryptographic systems would become insecure. Thus, the breakability of assumptions is often tied not only to hardware but also to the progression of theoretical knowledge.

- **Provable Security and Its Limitations:** The concept of provable security seeks to formalize cryptographic assumptions. In this framework, a cryptographic scheme is proven secure by reducing its security to a known hard problem. If breaking the scheme would require solving a problem assumed to be hard, the scheme is considered secure under that assumption. However, this does not mean absolute security. If the assumption is wrong, or if the reduction is loose or flawed, the system's security is jeopardized. Furthermore, real-world attacks often bypass theoretical models by targeting implementation weaknesses rather than the underlying mathematical assumptions.
- **Real-World Challenges to Assumptions:** The practical security of cryptographic systems depends not only on theoretical assumptions but also on how they are implemented. Side-channel attacks, such as power analysis and timing attacks, exploit physical characteristics of computing devices to recover secret information. Poor random number generation, flawed key management, and software bugs can also invalidate cryptographic protections, regardless of the underlying assumptions. Therefore, the strength of cryptographic assumptions must be evaluated in conjunction with practical considerations to ensure overall system security.
- **Importance of Cryptographic Agility:** Given the evolving nature of threats, cryptographic agility has become an essential design principle. This refers to the ability of a system to switch to stronger algorithms or larger key sizes as needed. For example, the transition from SHA-1 to SHA-2 and later to SHA-3 was driven by the discovery of collision vulnerabilities. Similarly, systems must be prepared to adopt post-quantum cryptographic algorithms in anticipation of quantum threats. Cryptographic agility allows for timely responses to broken assumptions without requiring a complete overhaul of the infrastructure.
- **Standardization and Community Response:** Organizations such as NIST, ISO, and IETF play a crucial role in managing cryptographic

standards and evaluating assumptions. NIST's Post-Quantum Cryptography Standardization Project is a prominent example of proactive planning for future threats. These organizations engage the global research community to rigorously analyze candidate algorithms, publish guidelines, and update standards based on emerging vulnerabilities. This collaborative process ensures that cryptographic assumptions are continuously tested and refined in response to new evidence.

- **Ethical and Policy Considerations:** The breakability of cryptographic assumptions also raises ethical and policy-related concerns. If a government or organization discovers a method to break a widely used cryptographic system but keeps it secret, it could compromise the security and privacy of countless users. This scenario creates a dilemma between national security and public trust. Transparency, responsible disclosure, and ethical research practices are essential to maintaining the integrity of cryptographic systems. Policymakers must balance the need for surveillance capabilities with the public's right to secure communication.

2.3 SECURITY LEVELS AND NOTIONS IN PQC

The field of post-quantum cryptography (PQC) has emerged in response to the impending threat quantum computers pose to traditional cryptographic systems. Unlike classical cryptography, which relies on mathematical problems like integer factorization or discrete logarithms, PQC utilizes algorithms resistant to quantum attacks. Security levels in PQC are designed to ensure protection against both classical and quantum adversaries, requiring rigorous analysis of computational hardness assumptions. The National Institute of Standards and Technology (NIST) has been instrumental in standardizing PQC algorithms, categorizing them based on security strength, performance, and resistance to quantum attacks.

<table>
<tr><th>PQC Algorithm</th><th>Status</th><th>Type</th><th>PKE/KEM vs. Signature</th></tr>
<tr><td>CRYSTALS-Kyber</td><td rowspan="4">Standard</td><td rowspan="3">Lattice</td><td>PEK/KEM</td></tr>
<tr><td>CRYSTALS-Dilithium</td><td rowspan="3">Signature</td></tr>
<tr><td>FALCON</td></tr>
<tr><td>SPHINCS$^+$</td><td>Hash</td></tr>
<tr><td>BIKE</td><td rowspan="3">Round 4</td><td rowspan="3">Code</td><td rowspan="4">PKE/KEM</td></tr>
<tr><td>Classic McEliece</td></tr>
<tr><td>HQC</td></tr>
<tr><td>SIKE</td><td>Broken</td><td>Isogeny</td></tr>
</table>

Fig: Family and Security Levels of PQC Algorithms

Post-quantum cryptography represents a fundamental shift in cryptographic security, necessitating new algorithms, larger key sizes, and stronger security notions. As quantum computing advances, the adoption of NIST-standardized PQC algorithms will be crucial for securing communications, financial systems, and critical infrastructure. While challenges remain in implementation and performance, ongoing research and standardization efforts aim to ensure a smooth transition to a quantum-resistant cryptographic future. Organizations must begin preparing now by evaluating PQC solutions, updating cryptographic protocols, and investing in quantum-safe security practices. Research continues to improve PQC security and efficiency:

- **Hybrid Cryptography**: Combining classical and PQC algorithms to ensure backward compatibility while transitioning to quantum-resistant systems.
- **Advanced Lattice-Based Schemes**: Exploring Ring-LWE and Module-LWE variants for better performance.
- **Zero-Knowledge Proofs**: Post-quantum secure ZKPs are being developed for privacy-preserving applications.
- **Hardware Acceleration**: Developing ASICs and FPGAs optimized for PQC operations to mitigate performance bottlenecks

2.3.1 Defining Security Levels in PQC

Security levels in post-quantum cryptography quantify the computational effort required to break an encryption scheme, typically measured in bits. A security level of *n* bits means that an attacker would need approximately 2^n operations to compromise the system. NIST classifies security levels into three primary categories: Level 1 (128-bit security), Level 3 (192-bit security), and Level 5 (256-bit security).

These levels correspond to the estimated resistance against attacks from both classical and quantum computers. For instance, AES-128 is considered secure against classical attacks but vulnerable to Grover's algorithm, reducing its effective security to 64 bits in a quantum setting. PQC algorithms must therefore compensate by increasing key sizes or leveraging mathematically harder problems to maintain equivalent security.

2.3.2 Security Notions in Post-Quantum Cryptography

Security notions define the formal guarantees that cryptographic algorithms must satisfy. In PQC, the most fundamental notions include:

- **IND-CPA (Indistinguishability under Chosen-Plaintext Attack)**: Ensures that an adversary cannot distinguish between ciphertexts of two chosen plaintexts.

- **IND-CCA2 (Indistinguishability under Adaptive Chosen-Ciphertext Attack):** A stronger notion where the adversary can request decryptions of chosen ciphertexts but still cannot break the scheme.
- **EUF-CMA (Existential Unforgeability under Chosen-Message Attack):** Critical for digital signatures, ensuring that an adversary cannot forge a valid signature for a new message.

These notions are adapted for quantum adversaries, requiring resistance against quantum-enhanced attacks such as Shor's algorithm, which can break RSA and ECC in polynomial time.

2.3.3 NIST's PQC Standardization and Security Benchmarks

NIST's PQC standardization process evaluates candidate algorithms based on their security, efficiency, and practicality. The selected algorithms fall into several families:

- **Lattice-Based Cryptography**: Includes Kyber (key encapsulation) and Dilithium (digital signatures), offering strong security guarantees based on the hardness of lattice problems like Learning With Errors (LWE).
- **Hash-Based Cryptography**: SPHINCS+ provides quantum-resistant signatures relying only on cryptographic hash functions.
- **Code-Based Cryptography**: Classic McEliece is based on error-correcting codes, resistant to quantum attacks due to the NP-hardness of decoding random linear codes.
- **Multivariate Cryptography**: Rainbow and other schemes rely on solving systems of multivariate quadratic equations, though some have been broken in recent years.

NIST mandates that approved algorithms meet at least Security Level 1 (128-bit equivalent) while recommending higher levels for long-term security.

2.3.4 Quantum Attack Models and Their Impact on Security

Post-quantum security must account for different quantum attack models:

- **Grover's Algorithm**: Reduces the security of symmetric key systems (e.g., AES) by a square root, necessitating larger key sizes (AES-256 for 128-bit post-quantum security).
- **Shor's Algorithm**: Breaks RSA, ECC, and Diffie-Hellman in polynomial time, rendering them obsolete in a quantum setting.
- **Hybrid Attacks:** Combine classical and quantum techniques to exploit weaknesses in PQC schemes.

Security proofs in PQC must demonstrate resistance against these attacks, often relying on worst-case hardness assumptions in computational complexity theory.

2.3.5 Challenges in Achieving Post-Quantum Security

Despite advancements, PQC faces several challenges:

- **Large Key Sizes:** Many PQC algorithms require significantly larger keys and signatures than classical systems, impacting storage and bandwidth.
- **Performance Overheads:** Lattice-based schemes, while secure, can be computationally intensive compared to RSA or ECC.
- **Side-Channel Attacks:** Physical implementations of PQC algorithms may leak information through timing or power analysis, requiring additional hardening.
- **Standardization Delays:** The transition from classical to post-quantum cryptography is complex, requiring updates to protocols, hardware, and legacy systems.

2.4 HYBRID CRYPTOGRAPHY: COMBINING CLASSICAL AND POST-QUANTUM SCHEMES

The advent of quantum computing presents both extraordinary opportunities and profound challenges to modern cryptography. While quantum machines promise breakthroughs in fields like drug discovery and optimization, they also threaten to break widely used encryption algorithms such as RSA and ECC. As organizations prepare for this shift, hybrid cryptography has emerged as a pragmatic solution—one that combines classical and post-quantum cryptographic schemes to ensure security today while future-proofing against quantum threats. This approach allows for a smoother transition, mitigating risks without requiring an immediate and complete overhaul of existing systems.

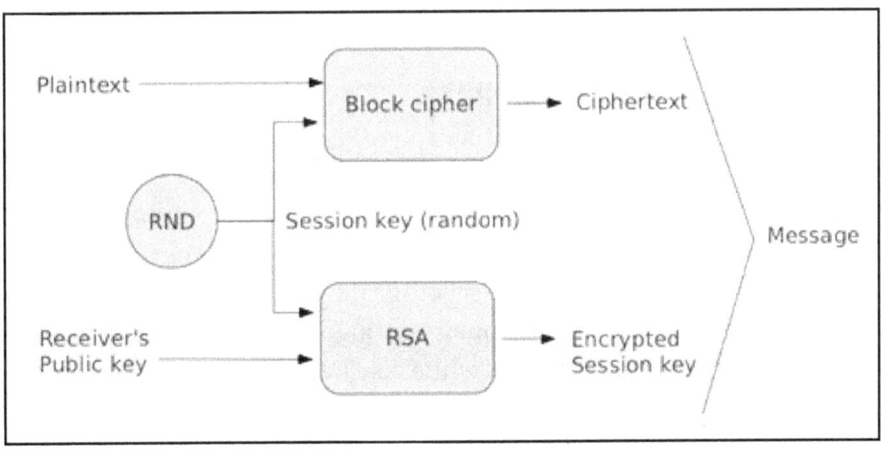

Fig: Hybrid Cryptosystem

Hybrid cryptography represents a pragmatic and strategic approach to addressing the challenges posed by quantum computing. By combining classical and post-quantum cryptographic schemes, organizations can enhance security, maintain compatibility, and prepare for future threats without undergoing abrupt and disruptive changes. While challenges such as implementation complexity and performance considerations exist, the benefits of hybrid systems—including gradual transition, enhanced security, and operational flexibility—far outweigh the drawbacks. As quantum computing advances, hybrid cryptography will be essential in safeguarding data, ensuring business continuity, and securing critical infrastructure. The time to adopt hybrid solutions is now, as they provide the necessary foundation for a secure and resilient digital future.

As quantum computing continues to evolve, hybrid cryptography will play an increasingly vital role in cybersecurity strategies. While the ultimate goal is a full transition to post-quantum cryptography, hybrid systems provide a necessary intermediate step. They allow organizations to adapt at their own pace, reducing risks while ensuring continuous protection. Over time, as PQC algorithms mature and quantum threats become more imminent, the reliance on classical cryptography will diminish, and hybrid systems will gradually shift toward full post-quantum implementations.

The development of hybrid cryptography also highlights the importance of collaboration among governments, industries, and academic institutions. Standardization bodies like NIST are working to define best practices for hybrid approaches, ensuring interoperability and security. Meanwhile, researchers continue to refine post-quantum algorithms to improve their efficiency and compatibility with classical methods. This collective effort will be crucial in shaping the future of cryptography, where hybrid solutions serve as a bridge to a fully quantum-resistant digital landscape.

2.4.1 Need for Hybrid Cryptography

The transition to post-quantum cryptography (PQC) is not a simple or immediate process. Many organizations rely on legacy systems that cannot be easily upgraded, and the standardization of PQC algorithms is still ongoing. Even as new quantum-resistant methods are developed, they may not yet be as optimized or widely tested as classical encryption. Hybrid cryptography addresses these challenges by integrating both classical and post-quantum algorithms, ensuring that systems remain secure against current threats while also being resilient to future quantum attacks.

One of the primary concerns in cryptography today is the "harvest now, decrypt later" strategy, where adversaries collect encrypted data with the intention of decrypting it once quantum computers become capable. Hybrid schemes mitigate this risk by ensuring that even if one layer of encryption is compromised, the other remains secure. This layered approach provides a safety net, allowing organizations to adopt PQC gradually while maintaining robust security in the interim. Additionally, hybrid cryptography helps bridge the gap between current infrastructure and future requirements, making the transition more manageable and less disruptive.

2.4.2 How Hybrid Cryptography Works

Hybrid cryptography combines classical and post-quantum algorithms in a way that leverages the strengths of both. For example, a hybrid system might use RSA or ECC for digital signatures alongside a lattice-based or hash-based PQC algorithm for encryption. This dual-layer approach ensures that even if one method is broken, the other remains intact, preserving security. The integration can occur at different levels, such as in key exchange protocols, digital signatures, or encryption schemes, depending on the specific use case and security requirements. One common implementation of hybrid cryptography is in Transport Layer Security (TLS), where classical and post-quantum algorithms are used together to secure communications. In such a setup, a classical algorithm might handle the initial handshake, while a PQC algorithm secures the data transmission. This ensures backward compatibility with existing systems while introducing quantum resistance. Similarly, hybrid digital signatures can combine classical and post-quantum methods to verify authenticity, providing an additional layer of security. The flexibility of hybrid approaches allows organizations to tailor their cryptographic strategies based on their risk tolerance, performance needs, and compliance requirements.

2.4.3 Advantages

One of the most significant benefits of hybrid cryptography is its ability to provide a seamless transition path to full PQC adoption. Rather than requiring an abrupt and potentially disruptive shift, hybrid systems allow organizations to introduce quantum-resistant elements incrementally. This gradual approach reduces implementation risks, minimizes compatibility issues, and provides time for thorough testing and optimization. It also allows businesses to maintain compliance with existing security standards while preparing for future regulations that may mandate PQC. Another advantage is enhanced security through redundancy. By using multiple cryptographic methods, hybrid systems make it significantly harder for attackers to compromise data. Even if a quantum computer eventually breaks one layer of encryption, the remaining classical or post-quantum layer would still protect the information. This defense-in-depth strategy is particularly valuable in high-stakes environments such as financial transactions, military communications, and critical infrastructure, where security breaches could have catastrophic consequences.

Additionally, hybrid cryptography helps address performance concerns associated with early PQC algorithms. Some post-quantum methods, while secure, may be computationally intensive or slower than classical alternatives. By combining them with efficient classical algorithms, hybrid systems can achieve a balance between security and performance. This is especially important for applications requiring real-time processing, such as online banking, IoT devices, and cloud computing, where latency and resource constraints are critical factors.

2.4.4 Challenges

Despite its advantages, hybrid cryptography is not without challenges. One of the primary concerns is the increased complexity of implementation. Combining multiple cryptographic schemes requires careful design to ensure that they work together seamlessly without introducing vulnerabilities. Poorly integrated hybrid systems could suffer from performance bottlenecks, compatibility issues, or even new attack vectors that exploit the interaction between classical and post-quantum methods.

Organizations must conduct rigorous testing and validation to ensure that hybrid solutions meet security and operational requirements. Another consideration is the potential for key management complications. Hybrid systems may require managing multiple key pairs—one for classical encryption and another for post-quantum methods—which can increase storage and computational overhead.

This could pose challenges for resource-constrained environments, such as embedded systems or legacy hardware. Additionally, the need to maintain both classical and post-quantum infrastructure could lead to higher operational costs, at least in the short term. Organizations must weigh these trade-offs against the long-term benefits of enhanced security and future readiness.

2.4.5 Real-World Applications

Hybrid cryptography is already being explored and implemented in various sectors where security and future-proofing are critical. Financial institutions, for example, are adopting hybrid approaches to secure online transactions and protect customer data. By combining classical encryption for immediate security with post-quantum methods for long-term resilience, banks can safeguard against both current cyber threats and future quantum attacks.

Similarly, government agencies and defense organizations are integrating hybrid cryptography into their communications and data storage systems to prevent espionage and ensure national security. The technology sector is also leveraging hybrid cryptography to secure cloud services, software updates, and IoT devices.

Companies like Google and Microsoft have already begun experimenting with hybrid TLS implementations to protect web traffic against quantum threats. In healthcare, hybrid encryption can secure electronic health records, ensuring that patient data remains confidential even as quantum computing advances. These real-world applications demonstrate the versatility and necessity of hybrid cryptography in preparing for the quantum era.

2.5 ROLE OF ENTROPY, RANDOMNESS, AND SIDE-CHANNEL RESISTANCE

2.5.1 Role of Entropy

Entropy, a fundamental concept derived from thermodynamics and information theory, plays a crucial role in modern cryptography and cybersecurity. In the context of cryptographic systems, entropy refers to the measure of randomness or unpredictability in data, which is essential for generating secure keys, ensuring robust encryption, and preventing attacks. Without sufficient entropy, cryptographic algorithms become vulnerable to exploitation, as predictable patterns can be exploited by adversaries to break encryption.

As cyber threats evolve and quantum computing introduces new challenges, understanding and harnessing entropy has become more critical than ever. Entropy is the unseen backbone of cryptographic security, ensuring that keys, random numbers, and authentication mechanisms remain unpredictable and resistant to attacks. As cyber threats grow more sophisticated and quantum computing introduces new vulnerabilities, the role of entropy becomes even more critical. From key generation to session security, high-quality entropy is essential for maintaining the integrity and confidentiality of digital communications.

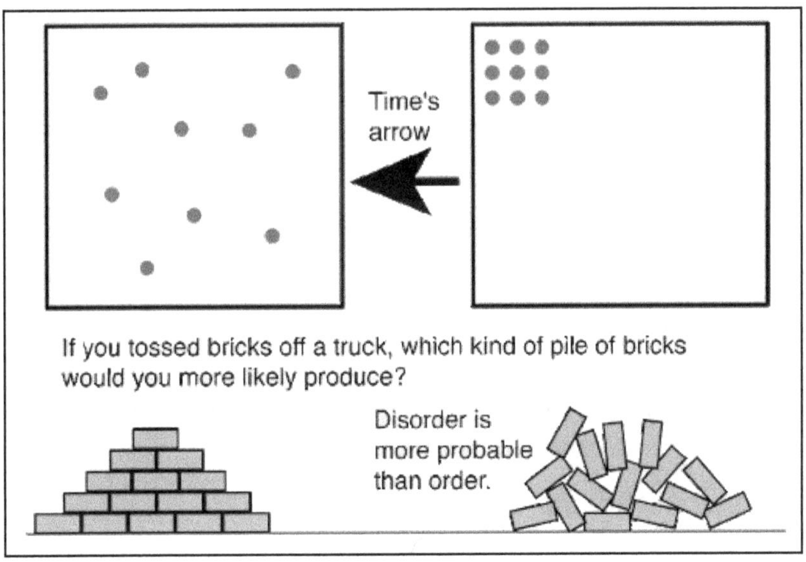

Fig: Entropy

Entropy in Cryptography

Entropy in cryptography is quantified as the degree of disorder or randomness present in a system. High entropy means that the output of a cryptographic process—such as a key or a random number—is highly unpredictable, making it resistant to brute-force attacks and cryptanalysis. The strength of encryption algorithms, digital signatures, and authentication mechanisms depends heavily on the quality of entropy used in their generation. If entropy is low, attackers can exploit patterns or biases to reconstruct keys or predict future outputs, compromising security. The concept of entropy is deeply rooted in information theory, where it measures the uncertainty of a random variable. In cryptography, this translates to the unpredictability of cryptographic keys.

For example, a truly random 256-bit key has an entropy of 256 bits, meaning there are 2^{256} possible combinations, making it computationally infeasible for an attacker to guess the correct key. However, if the entropy source is flawed—such as relying on predictable system states or weak random number generators—the effective security of the key diminishes. Thus, ensuring high-quality entropy is a cornerstone of secure cryptographic systems.

Entropy in Key Generation and Cryptographic Security

One of the most critical applications of entropy in cryptography is in the generation of cryptographic keys. Whether for symmetric encryption (e.g., AES) or asymmetric encryption (e.g., RSA, ECC), the security of these keys depends on their randomness. If an attacker can predict or reconstruct a key due to insufficient entropy, the entire encryption scheme can be compromised. This is particularly concerning in scenarios where keys are generated using deterministic algorithms or pseudo-random number generators (PRNGs) that lack true randomness.

To mitigate these risks, cryptographic systems rely on cryptographically secure pseudo-random number generators (CSPRNGs) and hardware-based entropy sources. CSPRNGs are designed to produce outputs that are computationally indistinguishable from true randomness, even if their initial seed is partially known.

Hardware-based entropy sources, such as those derived from physical phenomena like thermal noise or radioactive decay, provide higher-quality randomness by leveraging unpredictable natural processes. These methods ensure that keys generated for encryption, digital signatures, and authentication remain secure against both classical and quantum attacks.

Entropy in Random Number Generation

Random number generation (RNG) is another domain where entropy is indispensable. Cryptographic operations, including key generation, nonces, and initialization vectors, require high-quality randomness to prevent predictability. Weak RNGs have been the cause of numerous security breaches, such as the infamous Debian OpenSSL vulnerability in 2008, where a flawed RNG led to easily guessable encryption keys. Such incidents underscore the importance of robust entropy sources in cryptographic systems. Modern cryptographic systems often employ hybrid approaches to RNG, combining hardware-based entropy sources with algorithmic enhancements. For instance, a system might use environmental noise (e.g., mouse movements, disk latency) as an entropy source, which is then processed through a cryptographic hash function to produce a high-quality random seed. This seed can then be used in a CSPRNG to generate the necessary random numbers for cryptographic operations. By ensuring that the initial entropy pool is sufficiently random, these systems can resist attacks that attempt to exploit predictability in key generation or session tokens.

Entropy and Quantum Computing Threats

The rise of quantum computing introduces new challenges to cryptographic entropy. Quantum algorithms like Shor's algorithm can break classical public-key cryptography by solving the underlying mathematical problems (e.g., integer factorization, discrete logarithms) efficiently. However, even in the post-quantum era, entropy remains a critical defense mechanism. Post-quantum cryptographic algorithms, such as lattice-based or hash-based schemes, still rely on high-quality randomness for key generation and secure operations.

Quantum computers themselves can also be used to generate high-entropy random numbers through quantum random number generators (QRNGs). These devices exploit quantum phenomena, such as the behavior of photons or quantum superposition, to produce truly random outputs. QRNGs are considered among the most secure entropy sources available, as they are inherently unpredictable due to the fundamental principles of quantum mechanics. As quantum computing advances, the integration of QRNGs into cryptographic systems may become a standard practice for ensuring long-term security.

Entropy in Authentication and Session Security

Beyond encryption, entropy plays a vital role in authentication mechanisms and secure communication protocols. Session tokens, one-time passwords (OTPs), and cryptographic nonces all depend on high-entropy randomness to prevent replay attacks and session hijacking. For example, in TLS (Transport Layer Security), session keys are derived using random values to ensure that each session is unique and resistant to eavesdropping.

If these values are predictable due to poor entropy, attackers can exploit them to impersonate legitimate users or decrypt communications. Multi-factor authentication (MFA) systems also rely on entropy to generate secure tokens.

Hardware tokens, such as YubiKeys, use cryptographic operations seeded with high-entropy randomness to produce one-time codes that are resistant to brute-force attacks. Similarly, biometric authentication systems must ensure that the randomness used in their cryptographic processes is sufficiently strong to prevent spoofing or replay attacks. In all these cases, entropy is the invisible yet indispensable factor that underpins security.

Challenges in Maintaining High Entropy

Despite its importance, maintaining high entropy in cryptographic systems is not without challenges. One major issue is the availability of reliable entropy sources, particularly in constrained environments such as embedded systems or IoT devices. These systems often lack hardware-based randomness generators and must rely on software-based methods, which can be vulnerable to predictability if not properly implemented. Additionally, virtualized environments, such as cloud computing, face challenges in generating unique entropy across multiple instances, as shared resources can lead to entropy depletion or duplication.

Another challenge is the potential for entropy exhaustion, where a system's entropy pool is depleted faster than it can be replenished. This can occur in high-demand scenarios, such as during a denial-of-service attack where an attacker repeatedly requests cryptographic operations to drain the system's randomness. To counter this, modern operating systems implement mechanisms to replenish entropy pools dynamically, often using a combination of hardware and software-based sources. However, ensuring a consistent and secure supply of entropy remains an ongoing concern in cryptographic system design.

Best Practices for Ensuring Cryptographic Entropy

To mitigate the risks associated with insufficient entropy, several best practices should be followed. First, cryptographic systems should prioritize the use of hardware-based entropy sources whenever possible, as these provide the highest level of unpredictability. In environments where hardware sources are unavailable, software-based methods should be carefully designed to gather entropy from multiple unpredictable system events, such as user input timings or network latency.

Second, cryptographic libraries and protocols should be regularly audited to ensure that their entropy sources and RNG implementations are secure. Open-source projects, such as OpenSSL and LibreSSL, have faced scrutiny in the past for weaknesses in their randomness generation, highlighting the need for continuous evaluation. Additionally, systems should implement mechanisms to detect and respond to entropy depletion, such as blocking operations when randomness is insufficient or using cryptographic stretching techniques to extend limited entropy.

Finally, as quantum computing advances, the adoption of quantum-resistant entropy sources, such as QRNGs, should be considered for high-security applications. These devices can provide provably random outputs that are resistant to both classical and quantum attacks, ensuring long-term cryptographic security. By integrating these best practices, organizations can strengthen their cryptographic systems against evolving threats.

2.5.2 Randomness

Randomness is a fundamental pillar of modern cryptography and cybersecurity, serving as the bedrock upon which secure encryption, authentication, and digital signatures are built. Without true randomness, cryptographic systems become vulnerable to prediction, manipulation, and exploitation by adversaries. The concept of randomness extends beyond mere unpredictability—it encompasses the generation of numbers, keys, and tokens that must resist analysis, reverse engineering, and brute-force attacks.

As computing power grows and quantum computing emerges as a potential threat, the importance of high-quality randomness becomes even more pronounced. Randomness is the cornerstone of cryptographic security, ensuring that keys, tokens, and encryption processes remain unpredictable and resistant to attacks. From key generation to secure communication protocols, high-quality randomness is essential for maintaining the integrity and confidentiality of digital systems. As computing power grows and quantum computing introduces new vulnerabilities, the importance of robust randomness becomes even more critical.

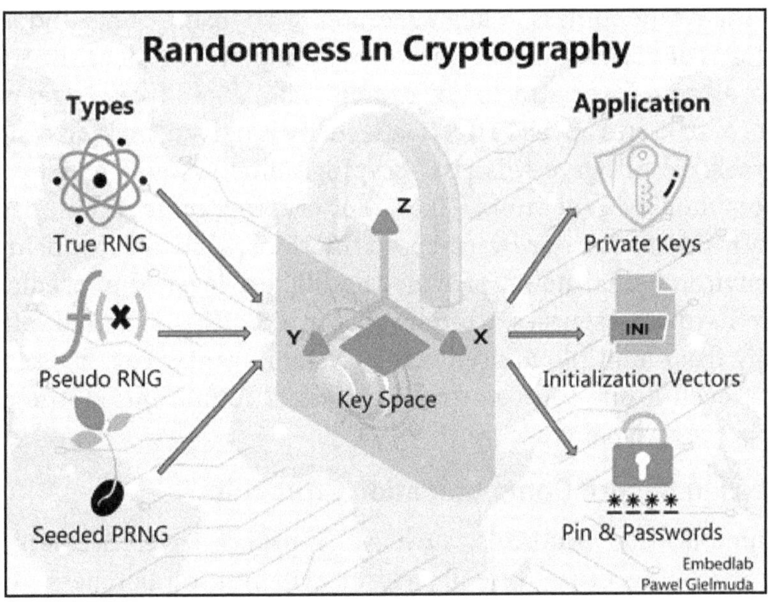

Fig: Randomness in Cryptography

Nature of Randomness in Cryptography

In cryptography, randomness refers to the property of being statistically unpredictable, meaning that the output of a random number generator cannot be guessed or reproduced without knowledge of the initial conditions. True randomness, often called entropy, is derived from unpredictable physical processes, such as atmospheric noise, thermal fluctuations, or quantum phenomena. However, in computational systems, true randomness is often supplemented or replaced by pseudo-randomness—algorithmic processes that produce sequences appearing random but are deterministic in nature.

The distinction between true randomness and pseudo-randomness is crucial in cryptography. True randomness is ideal for generating cryptographic keys because it eliminates any possibility of prediction. Pseudo-random number generators (PRNGs), while efficient, rely on a seed value, which, if compromised, could allow an attacker to reconstruct the entire sequence. Cryptographically secure pseudo-random number generators (CSPRNGs) address this by using cryptographic functions to ensure that even if part of the sequence is known, the rest remains unpredictable. This balance between efficiency and security makes CSPRNGs a practical choice for many cryptographic applications.

Role of Randomness in Key Generation

One of the most critical applications of randomness in cryptography is the generation of cryptographic keys. Whether for symmetric encryption (e.g., AES) or asymmetric encryption (e.g., RSA, ECC), the security of these keys depends entirely on their randomness. A key generated with insufficient randomness may exhibit patterns or biases that an attacker could exploit to reduce the search space required for a brute-force attack. For example, if a 256-bit key is generated with only 128 bits of true entropy, its effective security is halved, making it significantly easier to crack. To mitigate this risk, cryptographic systems employ hardware-based random number generators (HRNGs) or cryptographic techniques to ensure high-quality randomness. Hardware-based solutions, such as those using quantum effects or environmental noise, provide the highest level of unpredictability. In cases where hardware sources are unavailable, CSPRNGs are used, often seeded with entropy from multiple unpredictable system events. This layered approach ensures that even if one source of randomness is compromised, the overall key generation process remains secure.

Randomness in Secure Communication Protocols

Secure communication protocols, such as Transport Layer Security (TLS) and Internet Protocol Security (IPsec), rely heavily on randomness to establish secure sessions. During the handshake process, random values are exchanged to generate session keys that encrypt the communication. If these random values are predictable, an attacker could potentially decrypt the session or impersonate one of the parties. This vulnerability was famously exploited in the "Logjam" attack, where weak randomness in Diffie-Hellman key exchanges allowed attackers to downgrade encryption strength. To prevent such attacks, modern cryptographic protocols mandate the use of high-entropy randomness in key exchange mechanisms. For instance, TLS 1.3 requires that both the client and server contribute random values to the key derivation process, ensuring that even if one party's randomness is compromised, the session remains secure. Additionally, protocols often incorporate forward secrecy, where session keys are derived from ephemeral random values that are discarded after use, preventing retrospective decryption even if long-term keys are compromised.

Randomness in Authentication and Digital Signatures

Authentication mechanisms, including digital signatures and one-time passwords (OTPs), also depend on randomness to ensure security. Digital signatures, which verify the authenticity and integrity of messages, require randomness in the signing process to prevent signature forgery. For example, in the Digital Signature Algorithm (DSA), a random nonce is used in the signing process; if this nonce is reused or predictable, an attacker could recover the private key. This vulnerability was demonstrated in the Sony PlayStation 3 hack, where a reused nonce allowed attackers to extract the console's cryptographic keys.

Similarly, OTPs and session tokens rely on randomness to prevent replay attacks and brute-force guessing. If these tokens are generated with insufficient randomness, an attacker could predict future tokens or reuse intercepted ones. To counter this, modern authentication systems use cryptographically secure random number generators to produce tokens with high entropy, ensuring that each token is unique and unpredictable. This principle extends to multi-factor authentication (MFA), where randomness in token generation is essential to prevent unauthorized access.

Challenges in Generating True Randomness

Despite its importance, generating true randomness in computational systems presents significant challenges. One major issue is the reliance on deterministic processes in software-based random number generation. While CSPRNGs can produce sequences that appear random, their security ultimately depends on the quality of the initial seed. If the seed is predictable or derived from a weak entropy source, the entire sequence becomes vulnerable. This was evident in the Debian OpenSSL vulnerability, where a flawed random number generator led to easily guessable encryption keys. Another challenge is the availability of high-quality entropy sources in constrained environments, such as embedded systems or IoT devices. These systems often lack hardware-based randomness generators and must rely on software-based methods, which can be less secure. Additionally, virtualized environments, such as cloud computing, face difficulties in generating unique entropy across multiple instances, as shared resources can lead to entropy depletion or duplication. To address these challenges, modern systems often combine multiple entropy sources, such as hardware noise, user input timings, and cryptographic stretching techniques, to ensure a robust supply of randomness.

Quantum Computing and the Future of Randomness

The advent of quantum computing introduces both opportunities and challenges for randomness in cryptography. On one hand, quantum computers can break classical encryption algorithms, such as RSA and ECC, using Shor's algorithm, necessitating the adoption of post-quantum cryptographic methods. On the other hand, quantum mechanics provides a new source of true randomness through quantum random number generators (QRNGs). These devices leverage quantum phenomena, such as photon polarization or quantum superposition, to produce randomness that is fundamentally unpredictable.

QRNGs are considered among the most secure entropy sources available, as they are resistant to both classical and quantum-based attacks. As quantum computing advances, the integration of QRNGs into cryptographic systems may become a standard practice for high-security applications. However, the widespread adoption of QRNGs faces challenges, including cost, scalability, and integration with existing infrastructure. Despite these hurdles, the potential for quantum-enhanced randomness represents a significant advancement in cryptographic security.

Best Practices for Ensuring Cryptographic Randomness

To ensure the security of cryptographic systems, several best practices should be followed regarding randomness. First, systems should prioritize the use of hardware-based entropy sources, such as HRNGs or QRNGs, whenever possible. These sources provide the highest level of unpredictability and are resistant to algorithmic prediction. In environments where hardware sources are unavailable, software-based methods should be carefully designed to gather entropy from multiple unpredictable system events, such as network latency or disk read times.

Second, cryptographic libraries and protocols should be regularly audited to ensure that their random number generation processes are secure. Open-source projects, such as OpenSSL and LibreSSL, have faced scrutiny in the past for weaknesses in their randomness generation, highlighting the need for continuous evaluation. Additionally, systems should implement mechanisms to detect and respond to entropy depletion, such as blocking operations when randomness is insufficient or using cryptographic stretching techniques to extend limited entropy.

Finally, as quantum computing advances, organizations should consider the adoption of post-quantum cryptographic algorithms and quantum-resistant entropy sources. These measures will help future-proof cryptographic systems against emerging threats while maintaining the highest standards of randomness and security.

2.5.3 Side-Channel Resistance

In the realm of cryptography, security is often evaluated based on mathematical rigor and the theoretical difficulty of breaking encryption schemes. However, real-world implementations of cryptographic algorithms often exhibit unintended behaviors that can be exploited. These unintentional leaks are known as side channels, and the ability of a system to prevent or withstand such exploitation is referred to as side-channel resistance. Side-channel attacks do not target the theoretical weakness of an algorithm but exploit the physical characteristics and behavior of a device performing cryptographic computations. These include measurements of power consumption, electromagnetic emissions, timing variations, acoustic signals, and even cache access patterns.

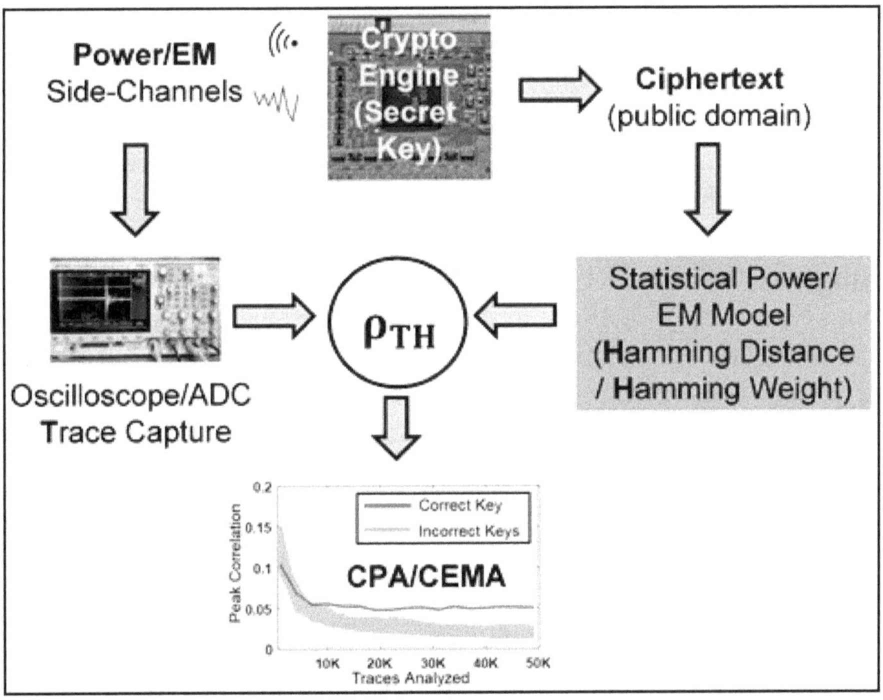

Fig: Electromagnetic and Power Side-Channel Analysis

Side-channel resistance is a crucial aspect of modern cryptographic security that ensures systems do not leak sensitive information through physical or behavioral characteristics. While cryptographic algorithms may be mathematically sound, poor implementation can lead to devastating real-world breaches. With the proliferation of connected devices and increasing sophistication of attackers, side-channel resistance must be treated as a core component of cryptographic design. It demands a holistic approach involving careful programming, hardware engineering, formal verification, and continuous testing. As the digital world grows more complex, the resilience of cryptographic systems against side-channel attacks will remain a defining challenge in cybersecurity.

Ongoing research in side-channel resistance focuses on creating lightweight, cost-effective, and formally verified countermeasures. Formal verification tools now assist developers in checking whether their software maintains constant-time properties or whether masking has been applied correctly. The field is also exploring machine learning techniques that simulate side-channel attacks more effectively, helping to predict and close vulnerabilities before devices reach the market. Furthermore, community-driven frameworks and repositories, such as the Test Vector Leakage Assessment (TVLA) and side-channel attack benchmarking suites, are enabling collaborative progress. Researchers are also increasingly working on side-channel resistance at the system level, recognizing that isolated software or hardware protections may not be enough in interconnected devices.

Side-channel attacks emerged as a powerful category of attacks in the late 1990s, especially when researchers demonstrated that even widely trusted encryption algorithms like RSA could be compromised using power analysis or timing attacks. Unlike brute-force attacks or cryptanalytic methods that require an attacker to solve complex mathematical problems, side-channel attacks derive information from the observable physical behavior of a system. For example, a timing attack takes advantage of the fact that certain operations take longer to perform depending on the input or the state of the key, thereby allowing the attacker to infer bits of the secret key. Similarly, power analysis examines fluctuations in a device's power consumption during cryptographic operations to extract key material. These types of attacks are often passive, making them difficult to detect and extremely dangerous.

Importance

As digital security becomes increasingly vital in sectors such as finance, defense, healthcare, and critical infrastructure, the need for side-channel resistance becomes more urgent. Devices such as smartcards, IoT sensors, contactless payment systems, and mobile phones are all vulnerable to side-channel attacks because they process sensitive information in potentially hostile environments. For example, a smartcard used for secure identification might be probed using a simple oscilloscope to measure power consumption and recover cryptographic keys. In such settings, it is not enough for a cryptographic algorithm to be mathematically secure; the device must also be implemented in a way that resists physical observation. Without side-channel resistance, even the strongest cryptographic systems can be rendered ineffective.

Types

Side-channel resistance must address a range of potential leaks. Timing attacks exploit differences in execution time that correlate with secret data. For instance, if a certain branch of code is only executed when a key bit is one, and that branch takes longer to execute, an attacker measuring response times can infer the value of that key bit. Power analysis attacks come in two main forms: simple power analysis (SPA) and differential power analysis (DPA).

SPA uses visual inspection of power traces to detect patterns, while DPA applies statistical analysis to multiple traces to isolate the influence of secret keys. Electromagnetic attacks involve capturing electromagnetic emissions from the device during operation. These emissions can reveal information similar to power traces but may be easier to collect remotely. Cache timing attacks exploit variations in memory access times, particularly in systems that use shared resources like CPUs. Even sound-based attacks have been demonstrated, where acoustic signals from keyboards or CPUs under load have been used to infer cryptographic keys.

Strategies for Building Side-Channel Resistance

Achieving side-channel resistance requires a multi-layered approach that combines software techniques, hardware design, and system-level policies. One of the primary methods for mitigating timing attacks is constant-time programming. This means ensuring that the program takes the same amount of time to execute regardless of input values or secret data. For example, cryptographic operations should not include branches or loops that depend on secret keys.

Masking is another technique where intermediate values in a cryptographic computation are randomized so that the actual values being manipulated are not directly related to the secret key. In DPA-resistant implementations, each operation on a sensitive variable is combined with a randomly generated mask, which is later removed at the end of the computation. Hardware countermeasures include noise generation, voltage balancing, and electromagnetic shielding to reduce the signal-to-noise ratio of measurable emissions. Some devices use dual-rail logic, which balances the power consumption of logic gates so that no useful information can be inferred from power traces.

Challenges

While the theory behind side-channel resistance is well understood, implementing effective defenses in practice is a significant challenge. First, many countermeasures introduce performance and cost overheads. Masking and constant-time programming often slow down cryptographic operations and require additional memory or processor resources. In constrained environments such as embedded systems or smartcards, these trade-offs can be difficult to manage. Second, achieving complete coverage of all possible leaks is nearly impossible. An implementation may be protected against power analysis but still vulnerable to electromagnetic leakage. Furthermore, some countermeasures themselves can introduce new vulnerabilities if not implemented correctly. For instance, an incorrect masking scheme can leak the mask used, effectively nullifying the protection. The complexity of modern systems and the difficulty of verifying complete security make side-channel resistance a persistent and evolving problem.

Testing and Evaluation

Verifying side-channel resistance requires rigorous testing and evaluation. This includes both theoretical analysis and practical experiments under controlled conditions. Security labs and certification bodies such as Common Criteria and EMVCo have established standards and procedures for evaluating side-channel resistance. Typically, devices are subjected to simulated attacks where power traces, electromagnetic emissions, or execution times are measured during cryptographic operations. Advanced equipment such as high-speed oscilloscopes, electromagnetic probes, and specialized test software are used to detect any statistical correlation between the observed data and secret keys. If such correlation is found, the device is considered vulnerable. The certification process is especially important for products used in critical applications, such as payment terminals, government ID cards, and military communication devices.

Side-Channel Resistance in the Age of Quantum Computing

While quantum computing is primarily discussed in the context of breaking classical cryptographic algorithms through Shor's or Grover's algorithms, its implications on side-channel resistance are also worth noting. Quantum side-channel attacks, though theoretical for now, may involve new forms of observation or manipulation that are not covered by current models. Additionally, the post-quantum cryptographic algorithms being developed today must also be evaluated for their side-channel resistance. Some of these algorithms, especially those based on lattice or multivariate polynomial assumptions, involve complex operations that may be difficult to implement securely without leaking side-channel information. Hence, side-channel resistance must be a core requirement in the transition to post-quantum cryptographic systems.

CHAPTER 3

Quantum Computing and the Threat Landscape

3.1 BASICS OF QUANTUM COMPUTING

Quantum computing is an emergent field of computer science and engineering that harnesses the unique qualities of quantum mechanics to solve problems beyond the ability of even the most powerful classical computers. The field of quantum computing includes a range of disciplines, including quantum hardware and quantum algorithms. While still in development, quantum technology will soon be able to solve complex problems that classical supercomputers can't solve (or can't solve fast enough).

By taking advantage of quantum physics, large-scale quantum computers would be able to tackle certain complex problems many times faster than modern classical machines. With a quantum computer, some problems that might take a classical computer thousands of years to solve might be solved in a matter of minutes or hours. Quantum mechanics, the study of physics at very small scales, reveals surprising fundamental natural principles. Quantum computers specifically harness these phenomena to access mathematical methods of solving problems not available with classical computing alone.

3.1.1 Practical Applications for Quantum Computing

In practice, quantum computers are expected to be broadly useful for two types of tasks: modeling the behavior of physical systems and identifying patterns and structures in information. Quantum mechanics is a bit like the operating system of the universe. A computer that uses quantum mechanical principles to process information has certain advantages in modeling physical systems. Therefore, quantum computing is of particular interest for chemistry and material science applications. For example, quantum computers might help researchers seeking useful molecules for pharmaceutical or engineering applications identify candidates more quickly and efficiently. Quantum computers can also process data by using mathematical techniques not accessible to classical computers. That means they can give structure to data and help discover patterns that classical algorithms alone might miss. In practice, this might be useful for applications ranging from biology (for example, protein folding) to finance.

Today, much of the work of quantum computing research involves searching for algorithms and applications within these broad categories of expected use. That is in addition to building the new technology itself.

3.1.2 Key Principles of Quantum Mechanics

When discussing quantum computers, it is important to understand that at the smallest scales, the universe behaves very differently from what we are used to in our day-to-day lives. Compared to what we learned in grade-school physics, the behaviors of quantum objects are often bizarre and counterintuitive.

Describing the behaviors of quantum particles presents a unique challenge. Most common-sense paradigms for the natural world lack the vocabulary to communicate the surprising behaviors of quantum particles. But quantum mechanics reveals how the universe really works. Quantum computers take advantage of quantum mechanics by replacing traditional binary bit circuits with quantum particles called quantum bits, or qubits. These particles behave differently from bits, exhibiting unique properties that can be described only with quantum mechanics.

- **Superposition:** A qubit itself isn't very useful. But it can place the quantum information it holds into a state of superposition, which represents a combination of all possible configurations of the qubit. Groups of qubits in superposition can create complex, multidimensional computational spaces. Complex problems can be represented in new ways in these spaces. When a quantum system is measured, its state collapses from a superposition of possibilities into a binary state, which can be registered like binary code as either a zero or a one.
- **Entanglement:** Entanglement is the ability of qubits to correlate their state with other qubits. Entangled systems are so intrinsically linked that when quantum processors measure a single entangled qubit, they can immediately determine information about other qubits in the entangled system.
- **Interference:** Interference is the engine of quantum computing. An environment of qubits placed into a state of collective superposition structures information in a way that looks like waves, with amplitudes associated with each outcome. These amplitudes become the probabilities of the outcomes of a measurement of the system. These waves can build on each other when many of them peak at a particular outcome or cancel each other out when peaks and troughs interact. Amplifying a probability or canceling out others are both forms of interference.
- **Decoherence:** Decoherence is the process in which a system in a quantum state collapses into a nonquantum state. It can be intentionally

triggered by measuring a quantum system or by other environmental factors (sometimes these factors trigger it unintentionally). Generally speaking, quantum computing requires avoiding and minimizing decoherence.

- **How the Principles Work Together:** To better understand quantum computing, consider that two surprising ideas are both true. The first is that objects that can be measured as having definite states—qubits in superposition with defined probability amplitudes—behave randomly. The second is that distant objects—in this case, entangled qubits—can still behave in ways that, though individually random, are strongly correlated. A computation on a quantum computer works by preparing a superposition of computational states. A quantum circuit, prepared by the user, uses operations to entangle qubits and generate interference patterns, as governed by a quantum algorithm. Many possible outcomes are canceled out through interference, while others are amplified. The amplified outcomes are the solutions to the computation.

3.1.3 How do Quantum Computers Work?

What are Qubits?

While classical computers rely on bits (zeros and ones) to store and process data, quantum computers process data differently by using quantum bits (qubits) in superposition. A qubit can behave like a bit and store either a zero or a one, but it can also be a weighted combination of zero and one at the same time. When qubits are combined, their superpositions can grow exponentially in complexity: two qubits can be in a superposition of the four possible 2-bit strings, three qubits can be in a superposition of the eight possible 3-bit strings, and so on. With 100 qubits, the range of possibilities is astronomical. Quantum algorithms work by manipulating information in a way inaccessible to classical computers, which can provide dramatic speed-ups for certain problems—especially when quantum computers and high-performance classical supercomputers work together.

Types of Qubits

Generally, qubits are created by manipulating and measuring systems that exhibit quantum mechanical behavior, such as superconducting circuits, photons, electrons, trapped ions and atoms. There are many different ways of making the qubits used in quantum computing today, with some better suited for different types of tasks. A few of the more common types of qubits in use are as follows:

- **Superconducting Qubits:** Made from superconducting materials operating at extremely low temperatures, these qubits are favored for their speed in performing computations and fine-tuned control.

- **Trapped Ion Qubits:** Trapped ion particles can also be used as qubits and are noted for long coherence times and high-fidelity measurements, but they are much slower than superconducting qubits.
- **Quantum Dots:** Quantum dots are small semiconductors that capture a single electron and use it as a qubit, offering promising potential for scalability and compatibility with existing semiconductor technology.
- **Photons:** Photons are individual light particles. They can be used to make qubits and send quantum information across long distances through optical fiber cables. They are being used in quantum communication and quantum cryptography.

Why are Qubits Useful?

Computers that use quantum bits have certain advantages over computers that use classical bits. Because qubits can hold a superposition and exhibit interference, a quantum computer that uses qubits approaches problems in ways different from classical computers. As a helpful analogy for understanding how quantum computers use qubits to solve complicated problems, imagine you are standing in the center of a complicated maze.

To escape the maze, a traditional classical computing approach would be to "brute force" the problem, trying every possible combination of paths to find the exit. This kind of computer would use bits to explore new paths and remember which ones are dead ends. A quantum computer might derive the correct path without needing to test all the bad paths, as if it has a bird's-eye view of the maze.

However, qubits don't test multiple paths at once. Instead, quantum computers measure the probability amplitudes of qubits to determine an outcome. These amplitudes function like waves, overlapping and interfering with each other. When asynchronous waves overlap, it effectively eliminates possible solutions to complex problems, and the realized coherent wave or waves present a correct solution.

3.1.4 Quantum Computing Components

An IBM quantum processor is a wafer not much bigger than the silicon chips found in a laptop. However, modern quantum hardware systems (used to keep the instruments at an ultracold temperature) and the extra room-temperature electronic components to control the system and process quantum data are about the size of an average car. While the large footprint of a complete quantum hardware system makes most quantum computers anything but portable, researchers and computer scientists are still able to access off-site quantum computing capabilities through cloud computing. The main hardware components of a quantum computer are as follows.

Quantum Processors

Composed of qubits laid out in various configurations to allow for communication, quantum chips—also known as the quantum data plane—act as the brain of the quantum computer. As the core component in a quantum computer, a quantum processor contains the system's physical qubits and the structures required to hold them in place. Quantum processing units (QPUs) include the quantum chip, control electronics and classical compute hardware required for input and output.

Superconductors

Your desktop computer likely uses a fan to get cold enough to work. Quantum processors need to be very cold—about a hundredth of a degree above absolute zero—to minimize noise and avoid decoherence in order to retain their quantum states. This ultralow temperature is achieved with supercooled superfluids. At these temperatures, certain materials exhibit an important quantum mechanical effect: electrons move through them without resistance. This effect makes them superconductors. When materials become superconductors, their electrons match up, forming Cooper pairs. These pairs can carry a charge across barriers, or insulators, through a process known as quantum tunneling. Two superconductors placed on either side of an insulator form a Josephson junction, a crucial piece of quantum computing hardware.

Control

Quantum computers use circuits with capacitors and Josephson junctions as superconducting qubits. By firing microwave photons at these qubits, we can control their behavior and get them to hold, change and read out individual units of quantum information.

Quantum Software

Research continues improving quantum hardware components, but that's only one half of the equation. The crux of users' discovery of quantum advantage will be a highly performant and stable quantum software stack to enable the next generation of quantum algorithms. In 2024, IBM introduced the first stable version of the Qiskit open source software development kit (SDK), Qiskit SDK 1.x. With over 600,000 registered users and 700 global universities that use it to develop quantum computing classes, Qiskit has become the preferred software stack for quantum computing.

But Qiskit is more than just the world's most popular quantum development software to build and construct quantum circuits. We are redefining Qiskit to represent the full-stack software for quantum at IBM, extending the Qiskit SDK with middleware software and services to write, optimize and run programs on IBM Quantum systems—including new generative AI code-assistance tools.

3.1.5 Classical Computing versus Quantum Computing

Quantum computing is built on the principles of quantum mechanics, which describe how very small objects behave differently from large objects. But because quantum mechanics provides the foundational laws for our entire universe, on a very small level, every system is a quantum system.

For this reason, we can say that while conventional computers are also built on top of quantum systems, they fail to take full advantage of the quantum-mechanical properties during their calculations. Quantum computers are expected to take better advantage of quantum mechanics to conduct calculations that even high-performance computers cannot.

Classical Computer

From antiquated punch-card adders to modern supercomputers, traditional (or classical) computers essentially function in the same way. These machines generally perform calculations sequentially, storing data by using binary bits of information. Each bit represents either a 0 or 1. When combined into binary code and manipulated by using logic operations, we can use computers to create everything from simple operating systems to the most advanced supercomputing calculations.

Quantum Computer

Quantum computers, like classical computers, are problem-solving machines. But instead of bits, quantum computing uses qubits. Qubits are used to process data like traditional bits; however, by harnessing quantum phenomena, qubits have access to more complex mathematics for a different type of computation. This is due to quantum mechanical concepts known as superposition and interference, which were discussed earlier.

Difference between Quantum and Classical Computing

Classical computing:

- Used by common, multipurpose computers and devices.
- Processes information in bits with a discrete number of possible states, 0 or 1.

Quantum computing:

- Used by specialized quantum mechanics-based quantum computing hardware.
- Processes information in qubits as 0, 1 or a superposition of 0 and 1.
- Processes data with quantum logic by using interference to solve problems.

Quantum processors do not perform mathematical equations the same way classical computers do. Unlike classical computers that must compute every step of a complicated calculation, quantum circuits made from logical qubits can process complex problems more efficiently.

While traditional computers commonly provide singular answers, probabilistic quantum machines often provide ranges of possible answers. This range might make quantum computing seem less precise than traditional computation. However, for the kinds of incredibly complex problems quantum computers might soon solve, this way of computing might potentially save hundreds of thousands of years of traditional computation.

In practice, quantum computers and classical computers work together in combined workflows to solve problems. The most efficient methods distribute the parts of a computation that quantum computers are best at to quantum computing resources and the parts that classical computers are best at to classical computing resources.

Fully realized quantum computers working in concert with high-performance classical computers would be far superior to classical computers alone for certain kinds of problems like integer factorization. But quantum computing is not ideal for every (or even most) problems.

3.1.6 When is Quantum Computing Superior

For most kinds of tasks and problems, classical computers are expected to remain the best solution. But when scientists and engineers encounter certain highly complex problems, quantum computing comes into play. For these types of difficult calculations, even the most powerful classical supercomputers pale in comparison to quantum computing. That's because even the most powerful classical supercomputers are binary code-based machines reliant on 20th-century technology.

Complex problems are problems with lots of variables interacting in complicated ways. For example, modeling the behavior of individual atoms in a molecule is a complex problem because of all the different interactions between electrons. Identifying new physics in a supercollider is also a complex problem. There are some complex problems that we do not know how to solve with classical computers at any practical scale.

A classical computer might be great at difficult tasks like sorting through a large database of molecules. But it struggles to solve more complex problems, like simulating how those molecules behave. Today, if scientists want to know how a molecule behaves, they must synthesize it and experiment with it in the real world. If they want to know how a slight tweak would impact its behavior, they usually need to synthesize the new version and run their experiment all over again. This is an expensive, time-consuming process that impedes progress in fields as diverse as medicine and semiconductor design. A classical supercomputer might try to simulate molecular behavior with brute force by using its many processors to explore every possible way every part of the molecule might behave. But as it moves past the simplest, most straightforward molecules available, the supercomputer stalls. No classical computer is able to handle all the possible permutations of molecular behavior by using any known methods.

Quantum algorithms take a new approach to these sorts of complex problems by creating multidimensional computational spaces in which to run algorithms that behave much like these molecules themselves. This turns out to be a much more efficient way of solving complex problems like chemical simulations.

One way to think about this: Classical computers need to crunch the numbers to figure out how a molecule will behave. A quantum computer doesn't need to crunch the numbers. It can mimic the molecular system directly. Quantum algorithms can also process data in ways classical computers can't, offering new structure and insights.

3.1.7 Quantum Computing Use Cases

First theorized in the early 1980s, it wasn't until 1994 that mathematician Peter Shor published one of the first practical real-world applications for a hypothetical quantum machine. Shor's algorithm for integer factorization demonstrated how a quantum mechanical computer could potentially break the most advanced cryptography systems of the time—some of which are still used today. Shor's findings demonstrated a viable application for quantum systems, with dramatic implications for not just cybersecurity, but many other fields. Engineering firms, financial institutions and global shipping companies, among others, are exploring use cases where quantum computers might solve important problems in their fields. An explosion of benefits from quantum research and development is taking shape on the horizon. As quantum hardware scales and quantum algorithms advance, we can soon find new solutions to big, important problems like molecular simulation, energy infrastructure management and financial market modeling.

Quantum computers excel at solving certain complex problems with many variables. From the development of new drugs to advancements in semiconductor development and tackling complex energy challenges, quantum computing might hold the key to breakthroughs in several critical industries.

- **Pharmaceuticals:** Quantum computers capable of simulating molecular behavior and biochemical reactions could speed up the research and development of life-saving new drugs and medical treatments.

- **Chemistry:** For the same reasons quantum computers could impact medical research, they might also provide undiscovered solutions for mitigating dangerous or destructive chemical byproducts. Quantum computing could lead to improved catalysts that enable petrochemical alternatives or better processes for the carbon breakdown necessary for combating climate-threatening emissions.
- **Machine Learning:** As interest and investment in artificial intelligence (AI) and related fields like machine learning ramps up, researchers are pushing AI models to new extremes, testing the limits of our existing hardware and demanding tremendous energy consumption. There is some reason to think that quantum algorithms might be able to look at datasets in a new way, providing a speed-up for some machine learning problems.

3.1.8 Quantum Utility versus Quantum Advantage

While no longer simply theoretical, quantum computing is still under development. As scientists around the world strive to discover new techniques to improve the speed, power and efficiency of quantum machines, technology is approaching a turning point. We understand the evolution of useful quantum computing using the concepts of quantum advantage and quantum utility.

- **Quantum Utility:** Quantum utility refers to any quantum computation that provides reliable, accurate solutions to problems that are beyond the reach of brute-force classical computing quantum-machine simulators. Previously, these problems were accessible only to classical approximation methods—usually problem-specific approximation methods carefully crafted to exploit the unique structures of a specific problem. IBM first demonstrated quantum utility in 2023.
- **Quantum Advantage:** Broadly defined, the term quantum advantage describes a situation where quantum can provide a better, faster, or cheaper solution than all known classical methods. An algorithm that exhibits quantum advantage on a quantum computer should be able to deliver a significant, practical benefit beyond all known classical computing methods. IBM expects that the first quantum advantages should be realized by late 2026, if the quantum and high-performance computing communities work together.

3.1.9 Quantum Benchmarks

Because quantum computing now offers a viable alternative to classical approximation for certain problems, researchers say it is a useful tool for scientific exploration, or that it has utility. Quantum utility does not constitute a claim that quantum methods have achieved a proven speed-up over all known classical methods. This is a key difference from the concept of quantum advantage.

- **Layer Fidelity:** An extremely valuable benchmark, layer fidelity provides a way to encapsulate the entire quantum processor's ability to run circuits while revealing information about individual qubits, gates and crosstalk. By running the layer fidelity protocol, researchers can qualify the overall quantum device while also gaining access to granular performance and error information about individual components.
- **Quantum Processing Speed:** In addition to layer fidelity, IBM also defined a speed metric: circuit layer operations per second (CLOPS). Currently, CLOPS is a measure of how quickly processors can run quantum volume circuits in series, acting as a measure of holistic system speed, incorporating quantum and classical computing. Together, layer fidelity and CLOPS provide a new way to benchmark systems that's more meaningful to the people trying to improve and use quantum hardware. These metrics make it easier to compare systems to one another, to compare our systems to other architectures and to reflect performance gains across scales.
- **Circuit Depth:** Circuit depth is also an essential capability of a quantum processing unit. It is a measure of the number of parallel gate executions—the number of steps in a quantum circuit—that the processing unit can run before the qubits decohere. The greater the circuit depth, the more complex circuits the computer can run.

3.1.10 Quantum Challenges and how to Make Quantum Computers more Useful

Today, companies like IBM, Google, Microsoft, D-Wave, Rigetti Computing and more make real quantum hardware. Cutting-edge tools that were merely theoretical four decades ago are now available to hundreds of thousands of developers. Engineers are delivering ever-more-powerful superconducting quantum processors at regular intervals, alongside crucial advances in software and quantum-classical orchestration. This work drives toward the quantum computing speed and capacity necessary to change the world.

Now that the field has achieved quantum utility, researchers are hard at work to make state-of-the-art quantum computers even more useful. Researchers at IBM Quantum and elsewhere have identified some key challenges to improve upon quantum utility and potentially achieve quantum advantage:

- **Scaling Quantum Processors:** While qubit processors used in quantum computing have the potential to massively outperform bit-based processors, current quantum processors can support only a small number of potential qubits. As research progresses, IBM plans to introduce a quantum system with 200 logical qubits capable of running 100 million quantum gates by 2029, with a goal of 2,000 logical qubits capable of running 1 billion gates by 2033.

- **Scaling Quantum Hardware:** Although powerful, qubits are also quite error-prone, requiring large cooling systems capable of creating temperatures lower than outer space. Researchers are currently developing ways to scale qubits, electronics, infrastructure and software to reduce footprint, cost and energy usage.
- **Quantum Error Correction:** Decoherence, the process in which qubits fail to function properly and produce inaccurate results, is a major hurdle for any quantum system. Quantum error correction requires that we encode quantum information into more qubits than we would otherwise need. In 2024, IBM announced a landmark new error-correcting code about 10 times more efficient than prior methods. While there are challenges remaining to implementing quantum error correcting codes and performing computations on encoded quantum information, this new code marks a clear path toward running quantum circuits with a billion logic gates or more.
- **Quantum Algorithm Discovery:** Quantum advantage requires two components. The first is viable quantum circuits, and the second is a means to demonstrate that those quantum circuits are better than known classical methods at solving a quantum problem. Quantum algorithm discovery is what will take current quantum technologies from quantum utility to quantum advantage.
- **Quantum Software and Middleware:** Quantum computing for advantage requires a highly performant and stable software stack to write, optimize and execute quantum programs. Open-source and Python-based, IBM's Qiskit is by far the most widely-used quantum SDK in the world. It is useful for executions both on IBM's fleet of superconducting quantum computers and on systems that use alternative technologies, such as ions trapped in magnetic fields or quantum annealing.
- **Quantum-Centric Supercomputing:** For the foreseeable future, quantum computing will work in tandem with modern and future classical supercomputing to be useful. In response, quantum researchers are preparing for a world where classical supercomputers can use quantum circuits to help solve problems.

3.2 SHOR'S ALGORITHM AND ITS IMPACT ON RSA, ECC

Quantum computing represents a revolutionary leap in computation, exploiting the principles of quantum mechanics to process information in fundamentally new ways. Unlike classical computers that use bits (0s and 1s), quantum computers use qubits, which can exist in superpositions of states. This allows quantum machines to explore multiple computational paths simultaneously, dramatically increasing their potential power for certain classes of problems. One of the most profound developments in quantum computing came in 1994 when Peter Shor introduced a quantum algorithm that could factor large integers and compute discrete logarithms in polynomial time.

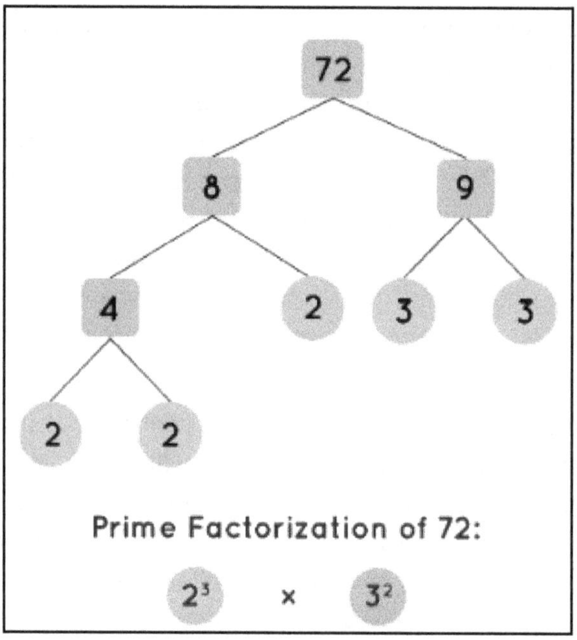

Fig: Shor's Algorithm and RSA Encryption

Shor's Algorithm stands as one of the most significant discoveries in quantum computing due to its potential to break foundational cryptographic systems like RSA and ECC. By enabling efficient solutions to factoring and discrete logarithm problems, it renders current public-key systems vulnerable in a future where scalable quantum computers are realized. The implications of this are vast, affecting every aspect of digital security from online communication to banking, identity verification, and national defense. In response, the global community has turned its attention toward post-quantum cryptography, aiming to develop and deploy new algorithms that can withstand quantum attacks. Though challenges in migration and standardization remain, the necessity of acting now is clear. The cryptographic world is at the threshold of a paradigm shift, and Shor's Algorithm is the catalyst demanding urgent and comprehensive change.

While Shor's Algorithm presents a theoretical quantum threat, practical implementation depends on the development of large-scale, fault-tolerant quantum computers. As of now, quantum hardware is still in its early stages, with limited qubit counts and high error rates. Researchers continue to explore the real-world feasibility of implementing Shor's Algorithm at scale, including improvements in qubit coherence, error correction, and gate fidelity. However, the timeline for achieving a quantum computer capable of breaking RSA or ECC remains uncertain. Some estimates suggest a 10 to 20-year horizon, while others believe it may happen sooner. Regardless of the exact timeline, the security community agrees that preparation must begin now to avoid catastrophic consequences when the quantum breakthrough arrives.

3.2.1 Mathematical Foundations

Shor's Algorithm is built upon quantum principles that enable it to solve the factoring and discrete logarithm problems exponentially faster than the best-known classical algorithms. For instance, factoring an nnn-bit number using classical methods such as the General Number Field Sieve takes sub-exponential time, which is still infeasible for very large numbers.

However, Shor's Algorithm can factor such numbers in polynomial time, specifically in time proportional to O(n3)O(n^3)O(n3), using quantum Fourier transforms and modular arithmetic. At its core, the algorithm works by converting the factoring problem into a problem of period finding, which quantum systems can solve efficiently using interference and entanglement. This breakthrough effectively renders RSA and ECC insecure in a future where large-scale quantum computers are available, because the security of these systems depends directly on the infeasibility of factoring and discrete logarithms.

3.2.2 RSA and the Threat of Quantum Factorization

The RSA cryptosystem, one of the most widely deployed public-key cryptographic methods, derives its security from the difficulty of factoring the product of two large prime numbers. In the RSA algorithm, a public key is used for encryption and a private key is used for decryption, with the assumption that deriving the private key from the public key is computationally impractical.

Classical factoring algorithms cannot break RSA unless the key size is small or poorly chosen. However, Shor's Algorithm breaks this assumption by making factoring feasible on a quantum computer. For example, a quantum computer with a sufficient number of stable qubits and low error rates could break a 2048-bit RSA key within hours, something that would take classical supercomputers billions of years. This potential renders all existing RSA-based communication vulnerable, including SSL/TLS protocols, digital signatures, and secure email systems.

3.2.3 Elliptic Curve Cryptography and Discrete Logarithms

Elliptic Curve Cryptography (ECC) is another widely adopted cryptographic scheme that offers similar security to RSA but with smaller key sizes, making it highly efficient for use in mobile devices, smart cards, and IoT systems. ECC is based on the difficulty of solving the elliptic curve discrete logarithm problem (ECDLP). In ECC, finding the private key from the public key requires computing the discrete logarithm on an elliptic curve, which is computationally infeasible using classical algorithms for large enough curves.

However, Shor's Algorithm can also solve discrete logarithm problems in polynomial time. This includes ECDLP, meaning that ECC is no more secure than RSA in a quantum computing context. While ECC has been promoted as a lightweight and efficient cryptographic alternative to RSA, its vulnerability to quantum attacks places it in the same jeopardy.

3.2.4 Wider Implications for Cryptographic Infrastructure

The impact of Shor's Algorithm extends beyond individual encryption schemes and threatens the very foundation of current public-key infrastructure. Protocols that rely on RSA or ECC for key exchange, authentication, and digital signatures would all be susceptible to compromise. This includes HTTPS, VPNs, secure emails, blockchain technologies, and digital identity systems. Moreover, any encrypted data that has been captured today could be decrypted in the future once a quantum computer becomes powerful enough, a concept known as "harvest now, decrypt later." The transition to post-quantum cryptography is therefore not just a theoretical need but a practical urgency. Sensitive information transmitted today using RSA or ECC may be vulnerable in the near future unless protected with quantum-resistant algorithms.

3.2.5 Race towards Post-Quantum Cryptography

In response to the quantum threat posed by Shor's Algorithm, cryptographic researchers have been developing post-quantum cryptography (PQC), which aims to create cryptographic systems that are secure even against quantum adversaries. These systems rely on hard problems that, to the best of current knowledge, cannot be solved efficiently by quantum algorithms. Examples include lattice-based cryptography, code-based cryptography, multivariate polynomial equations, and hash-based signatures.

Lattice-based systems, such as the Learning With Errors (LWE) problem, are considered among the most promising candidates due to their strong theoretical foundations and efficiency. The National Institute of Standards and Technology (NIST) has initiated a global competition to standardize post-quantum cryptographic algorithms, with several finalists already under evaluation. These new standards are expected to replace RSA and ECC in applications across government, industry, and consumer technologies.

3.2.6 Challenges in Transitioning from RSA and ECC

Transitioning from RSA and ECC to post-quantum cryptographic systems is a complex process that presents several challenges. One major concern is backward compatibility, as existing systems are deeply entrenched in the current public-key infrastructure. Replacing RSA and ECC involves updating protocols, software, hardware, and policies across a vast range of applications. Additionally, many of the proposed quantum-resistant algorithms have different performance characteristics compared to RSA or ECC. Some require more computational power, larger key sizes, or more memory, which may not be feasible for all environments. There are also concerns about the security assumptions underlying new algorithms, as they have not undergone the decades of scrutiny that RSA and ECC have endured. Nonetheless, the transition is unavoidable, and work is already underway to implement hybrid systems that combine classical and quantum-resistant algorithms during the migration period.

3.2.7 Quantum Readiness and Strategic Planning

Organizations, governments, and industry leaders must begin preparing for a quantum future by assessing their current cryptographic assets and planning for a phased migration. Quantum readiness involves identifying which systems rely on RSA or ECC, understanding their exposure to long-term confidentiality risks, and developing a roadmap to implement quantum-resistant solutions. Strategic planning includes engaging with standards bodies, testing new algorithms, training personnel, and updating legal and regulatory frameworks to accommodate the changes. The earlier these steps are taken, the smoother the transition will be, minimizing disruption and safeguarding data. In particular, critical infrastructure, national defense systems, and financial institutions must prioritize quantum readiness due to the sensitivity and longevity of the information they handle.

3.3 GROVER'S ALGORITHM AND ITS IMPACT ON SYMMETRIC CIPHERS

The advent of quantum computing has introduced both revolutionary possibilities and significant challenges to modern cryptography. Among the most consequential quantum algorithms is Grover's algorithm, which demonstrates the potential to disrupt symmetric encryption—the backbone of secure communications and data protection. Unlike Shor's algorithm, which threatens public-key cryptography by efficiently solving integer factorization and discrete logarithm problems, Grover's algorithm poses a different but equally profound risk by accelerating brute-force search operations. This capability could effectively halve the security strength of symmetric ciphers, forcing cryptographers and security professionals to reevaluate encryption standards and key lengths. Grover's algorithm represents a significant but manageable threat to symmetric cryptography, necessitating proactive adjustments to key lengths and cryptographic practices. While it does not render symmetric encryption obsolete, it underscores the need for increased key lengths and the adoption of hybrid cryptographic systems to maintain security in the quantum era. Organizations must prioritize cryptographic agility, research into quantum-resistant algorithms, and the potential integration of Quantum Key Distribution to future-proof their security infrastructures. By taking these steps, the resilience of symmetric encryption can be preserved, ensuring that data remains secure against both classical and quantum threats. The transition to quantum-resistant cryptography is not merely a technical challenge but a strategic imperative for safeguarding digital communications in the decades to come.

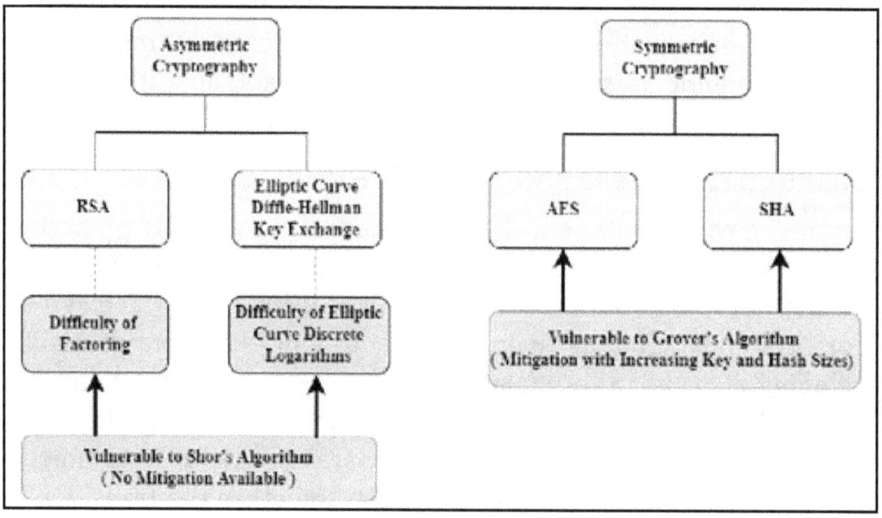

Fig: Effect of Shor's and Grover's Algorithms on Cryptography Algorithm

Grover's algorithm, developed by Lov Grover in 1996, is a quantum search algorithm that provides a quadratic speedup for unstructured search problems. In classical computing, searching an unsorted database of N elements requires, on average, N/2 operations. Grover's algorithm, however, can perform the same search in approximately $\sqrt{N}$ operations, offering a significant advantage in computational efficiency. While this speedup may seem modest compared to the exponential acceleration provided by Shor's algorithm, its implications for cryptography are profound, particularly for symmetric encryption schemes that rely on brute-force resistance.

The algorithm operates by leveraging quantum superposition and interference to amplify the probability of measuring the correct solution. It iteratively applies a sequence of quantum operations that invert the phase of the target state while preserving the amplitudes of other states. Through repeated applications, the probability of measuring the desired solution increases, eventually converging to near certainty. This process effectively reduces the search space, making it feasible to break symmetric encryption schemes that were previously considered secure against classical brute-force attacks.

3.3.1 Impact on Symmetric Cipher Security

Symmetric ciphers, such as the Advanced Encryption Standard (AES), rely on the computational infeasibility of brute-force attacks to ensure security. The strength of these ciphers is typically measured in terms of key length, with longer keys providing exponentially greater resistance to exhaustive search.

For example, a 128-bit AES key has 2^a possible combinations, making brute-force attacks impractical with classical computers. However, Grover's algorithm changes this calculus by reducing the effective security of symmetric keys. When applied to symmetric encryption, Grover's algorithm can theoretically reduce the time required to find a key from $O(2^n)$ to $O(2^n/^2)$, effectively halving the security strength.

This means that a 128-bit key, which would require an impractical amount of time to crack with classical methods, could be broken in a feasible timeframe with a sufficiently powerful quantum computer. To maintain the same level of security, cryptographic keys would need to be doubled in length. For instance, a 128-bit key would need to be upgraded to 256 bits to preserve its resistance against quantum-enhanced brute-force attacks.

3.3.2 Practical Implications for Cryptographic Standards

The threat posed by Grover's algorithm necessitates a reevaluation of current cryptographic standards and practices. While symmetric encryption remains more resilient to quantum attacks than public-key cryptography, the reduced security margin demands proactive measures. Organizations relying on symmetric ciphers must consider increasing key lengths to mitigate the risk of quantum brute-force attacks.

For example, AES-128, which is widely used today, may need to be replaced with AES-256 to maintain long-term security in a post-quantum world. However, simply increasing key lengths is not always straightforward. Some legacy systems may not support longer keys due to hardware or software limitations, requiring significant upgrades or replacements. Additionally, longer keys can introduce performance overhead, particularly in resource-constrained environments such as IoT devices or embedded systems. Balancing security and efficiency will be a critical consideration as organizations transition to quantum-resistant cryptographic practices.

3.3.3 Comparative Resilience of Symmetric vs. Asymmetric Cryptography

While Grover's algorithm presents a significant threat to symmetric encryption, its impact is less severe than that of Shor's algorithm on asymmetric cryptography. Shor's algorithm can break widely used public-key schemes like RSA and ECC in polynomial time, rendering them obsolete in the face of quantum computing. In contrast, Grover's algorithm only provides a quadratic speedup, meaning symmetric ciphers remain viable with appropriate key length adjustments. This relative resilience makes symmetric encryption a more robust choice for long-term security, provided that key lengths are sufficiently increased.

This distinction has led to a growing emphasis on symmetric cryptography in post-quantum security strategies. Hybrid cryptographic systems, which combine classical and post-quantum algorithms, often rely on symmetric encryption for bulk data protection while using quantum-resistant public-key schemes for key exchange. This approach leverages the strengths of both cryptographic paradigms, ensuring security against both classical and quantum threats. As quantum computing advances, symmetric encryption is likely to play an even more central role in secure communications and data storage.

3.3.4 Mitigation Strategies and Future-Proofing

To defend against the potential impact of Grover's algorithm, organizations must adopt a multi-faceted approach to cryptographic security. The most immediate step is to transition to longer key lengths for symmetric ciphers, such as moving from AES-128 to AES-256. This adjustment ensures that the effective security remains robust even when subjected to quantum-enhanced brute-force attacks. Additionally, cryptographic agility—the ability to update algorithms and key lengths dynamically—will be essential for adapting to evolving threats.

Another critical strategy is the adoption of hybrid cryptographic systems that combine classical and post-quantum algorithms. For example, a system might use a quantum-resistant key exchange protocol to establish a symmetric key, which is then used with a longer key length for encryption. This approach provides defense-in-depth, ensuring that even if one component is compromised, the overall system remains secure. Furthermore, ongoing research into quantum-resistant symmetric ciphers and cryptographic primitives will be vital for maintaining security in the long term.

3.3.5 Role of Quantum Key Distribution

Quantum Key Distribution (QKD) represents another promising avenue for securing symmetric encryption against quantum threats. QKD leverages the principles of quantum mechanics to enable two parties to generate a shared, secret key with information-theoretic security. Unlike classical key exchange methods, which rely on computational hardness assumptions, QKD provides security based on the laws of physics, making it resistant to both classical and quantum attacks.

While QKD is not a direct replacement for symmetric encryption, it offers a secure method for key distribution, which is often the most vulnerable part of cryptographic systems. By using QKD to establish symmetric keys, organizations can ensure that their encryption remains secure even in the presence of quantum computers. However, QKD faces practical challenges, including high implementation costs, limited range, and the need for specialized hardware. Despite these hurdles, ongoing advancements in QKD technology may make it a viable component of future cryptographic infrastructures.

3.4 ATTACK SCENARIOS AND RISK MODELING

The advent of quantum computing introduces transformative possibilities for computation while simultaneously posing unprecedented risks to existing cryptographic systems. Unlike classical computers that process information in binary bits, quantum computers leverage qubits capable of superposition and entanglement, enabling them to solve certain mathematical problems exponentially faster.

This capability threatens the foundation of modern encryption, particularly public-key cryptography, which secures digital communications, financial transactions, and government systems. Quantum computing represents a paradigm shift in both computational capability and cybersecurity risk. The potential for quantum attacks against classical cryptography demands urgent action, from adopting post-quantum algorithms to rethinking long-term data protection strategies. Risk modeling must evolve to address not only immediate cryptographic vulnerabilities but also the broader implications of quantum-enhanced cyber warfare.

The quantum threat landscape is evolving rapidly, requiring proactive measures from both public and private sectors. Governments must establish regulatory mandates for PQC adoption in critical industries, while enterprises should conduct comprehensive cryptographic inventories to identify vulnerable systems. Academic and industry partnerships are essential for advancing quantum-safe research and developing standardized best practices.

Continuous monitoring of quantum computing advancements ensures that risk models remain up-to-date as new attack vectors emerge. By integrating quantum risk management into broader cybersecurity strategies, organizations can navigate the transition to a post-quantum world with confidence, ensuring the confidentiality, integrity, and availability of digital systems in the face of unprecedented computational power.

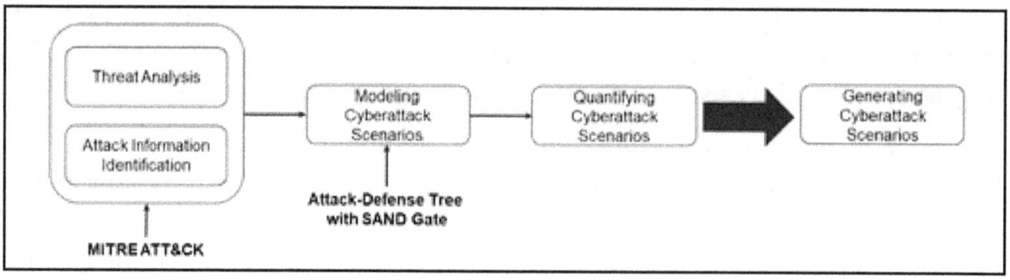

Fig: Attack Scenarios

3.4.1 Quantum Attack Vectors against Classical Cryptography

Quantum computers exploit specific algorithms that undermine classical cryptographic schemes. Shor's algorithm, one of the most well-known quantum algorithms, efficiently factors large integers and computes discrete logarithms, rendering RSA and elliptic curve cryptography (ECC) obsolete once sufficiently powerful quantum computers emerge. Grover's algorithm, while less destructive, provides a quadratic speedup for brute-force searches, effectively halving the security strength of symmetric encryption like AES. These attack vectors necessitate a reevaluation of cryptographic standards, as current encryption methods may no longer provide sufficient protection. Additionally, hybrid attacks combining classical and quantum techniques could exploit transitional vulnerabilities during the migration to post-quantum cryptography (PQC), making risk assessment a critical priority for enterprises and governments alike.

3.4.2 Harvest Now, Decrypt Later: A Looming Threat

One of the most concerning attack scenarios in the quantum threat landscape is the "Harvest Now, Decrypt Later" (HNDL) strategy. Adversaries with long-term objectives may already be collecting and storing encrypted data, anticipating future access to quantum decryption capabilities.

Government agencies, financial institutions, and healthcare providers handling highly sensitive information are particularly vulnerable to such attacks. Even if quantum computers capable of breaking current encryption are years away, the data stolen today could be decrypted retroactively, exposing trade secrets, personal identities, and classified communications. Proactive risk modeling must account for this delayed decryption threat, emphasizing the urgency of transitioning to quantum-resistant cryptographic solutions before large-scale quantum computers become operational.

3.4.3 Side-Channel Attacks in a Quantum Context

While much attention focuses on mathematical cryptanalysis, quantum computing also introduces new side-channel vulnerabilities. Quantum systems are highly sensitive to environmental noise, and adversaries could exploit physical weaknesses in quantum hardware to extract secret keys. Timing attacks, power analysis, and fault injection techniques may be adapted to target quantum cryptographic implementations. Furthermore, the nascent state of quantum error correction and noise-resistant quantum gates presents additional attack surfaces. Risk models must incorporate these physical-layer threats, ensuring that quantum-secure protocols are not only theoretically sound but also resilient against real-world implementation flaws. Research into quantum-safe hardware security modules (HSMs) and tamper-resistant quantum key distribution (QKD) systems is critical to mitigating these risks.

3.4.4 Post-Quantum Cryptographic Migration Risks

The transition from classical to post-quantum cryptography introduces its own set of risks and challenges. Many PQC algorithms, such as lattice-based and hash-based schemes, are still undergoing standardization and optimization. Early adopters may face interoperability issues, performance bottlenecks, and unforeseen vulnerabilities as these algorithms are tested at scale. Additionally, legacy systems embedded in critical infrastructure may resist easy upgrades, creating security gaps during the migration period. Risk modeling must account for transitional threats, including downgrade attacks where adversaries force systems to revert to weaker cryptographic protocols. A phased implementation strategy, combining hybrid cryptography (classical + PQC) with continuous vulnerability assessments, can help mitigate these risks while maintaining backward compatibility.

3.4.5 Quantum Computing in Offensive Cyber Operations

Beyond breaking encryption, quantum computing could revolutionize offensive cyber operations. Quantum-enhanced machine learning algorithms may accelerate password cracking, malware optimization, and network intrusion detection evasion. Adversarial quantum computing could also disrupt blockchain systems by undermining consensus mechanisms or enabling double-spending attacks. Nation-state actors and cybercriminal organizations are likely to invest in quantum capabilities, making it imperative for defensive strategies to evolve in tandem. Risk models should evaluate not just cryptographic vulnerabilities but also the broader implications of quantum-empowered cyber warfare, including AI-driven attacks and autonomous exploit development.

3.4.6 Risk Quantification and Threat Modeling Frameworks

To systematically assess quantum-related threats, organizations must adopt structured risk modeling frameworks. The National Institute of Standards and Technology (NIST) and the European Telecommunications Standards Institute (ETSI) have begun developing guidelines for quantum risk assessment, emphasizing cryptographic agility and zero-trust architectures. Key components of effective risk modeling include asset criticality analysis, adversary capability estimation, and impact severity scoring. Probabilistic models can help quantify the likelihood of quantum attacks based on projected advancements in quantum hardware. Additionally, red teaming exercises simulating quantum-enabled adversaries can identify weaknesses in existing security postures. These frameworks enable organizations to prioritize mitigation efforts, allocate resources efficiently, and develop contingency plans for quantum-related breaches.

3.4.7 Mitigation Strategies for Quantum Threats

Addressing quantum risks requires a multi-layered defense strategy. The immediate priority is the adoption of NIST-approved post-quantum cryptographic standards, such as CRYSTALS-Kyber for key exchange and CRYSTALS-Dilithium for digital signatures. For long-term data protection, enterprises should implement forward-secrecy protocols and quantum-secure key management solutions. Hybrid encryption models, which combine classical and PQC algorithms, provide interim protection during the transition period. Beyond cryptography, organizations should invest in quantum-resistant network architectures, including intrusion detection systems capable of identifying anomalous quantum computing activity. Collaboration with cybersecurity agencies, academic researchers, and industry consortia ensures access to the latest threat intelligence and mitigation techniques.

3.4.8 Role of Quantum Key Distribution (QKD)

Quantum Key Distribution (QKD) offers a theoretically unbreakable method for secure key exchange by leveraging the principles of quantum mechanics. Unlike classical key exchange protocols, QKD detects eavesdropping attempts through quantum state disturbances, ensuring immediate awareness of interception. While QKD is currently limited by range and infrastructure requirements, advancements in satellite-based quantum communication and trusted node networks are expanding its feasibility. Risk models should evaluate QKD as a complementary solution for high-security environments, such as government communications and financial transactions, where traditional key exchange methods may be vulnerable. However, practical challenges, including cost, deployment complexity, and integration with existing systems, must be carefully weighed against its security benefits.

3.5 QUANTUM-RESISTANT VS. QUANTUM-PROOF: TERMINOLOGY AND MISCONCEPTIONS

Quantum computing represents a revolutionary advancement in computational technology, harnessing the principles of quantum mechanics to solve problems that are currently intractable for classical computers. Unlike traditional systems that use bits as the smallest unit of information, quantum computers utilize qubits, which can exist in multiple states simultaneously thanks to superposition, and can be entangled with each other, enabling exponentially greater processing power. This unprecedented capability holds promise for breakthroughs across various fields, including cryptography, drug discovery, and complex system modeling.

However, it also introduces significant challenges to the existing digital security infrastructure, particularly in the realm of cryptography, where many current encryption algorithms rely on the difficulty of factoring large numbers or solving discrete logarithm problems—tasks that quantum algorithms like Shor's algorithm could potentially solve efficiently. As a result, the threat landscape is evolving rapidly, prompting researchers and security professionals to develop new cryptographic standards capable of resisting quantum attacks.

The distinction between quantum-resistant and quantum-proof cryptography becomes crucial in this context. Quantum-resistant (or post-quantum) algorithms are designed to withstand attacks from quantum computers but have not yet been proven to be unbreakable against all conceivable quantum threats; they are currently considered the best available defense while the technology matures. Quantum-proof, on the other hand, implies an absolute security guarantee against quantum attacks, often suggesting that the algorithms are proven to be unbreakable even by quantum computers, or that they employ fundamentally different principles that make quantum attacks infeasible.

The transition to quantum-resistant cryptography is already underway, with initiatives such as the National Institute of Standards and Technology (NIST) leading efforts to standardize post-quantum algorithms that can secure data against future quantum threats. This proactive approach aims to safeguard sensitive information, especially long-term data that could be vulnerable once powerful quantum computers become operational. Despite these efforts, the threat is not purely technical; it also encompasses logistical, economic, and geopolitical challenges. Governments and organizations must not only adopt new cryptographic standards but also ensure widespread implementation and transition from legacy systems, which may be deeply embedded in existing infrastructure.

The threat landscape continues to grow as quantum computing advances, with potential risks extending beyond cryptography to include the disruption of financial systems, communications, and national security. As such, understanding the difference between quantum-resistant and quantum-proof systems is essential for developing resilient security strategies. While quantum-resistant algorithms represent a significant step forward, their effectiveness depends on ongoing research, implementation, and testing to ensure they can withstand future quantum capabilities. On the other hand, the notion of quantum-proof security, although desirable, remains a theoretical ideal, emphasizing the importance of foundational research into new cryptographic paradigms that could inherently prevent quantum exploitation. Ultimately, the evolution of quantum computing and its impact on the threat landscape underscores the need for a proactive, layered approach to cybersecurity, combining technological innovation, standardization efforts, and international cooperation to mitigate vulnerabilities and ensure the integrity and confidentiality of digital information in the quantum era.

CHAPTER 4

Post-Quantum Cryptographic Algorithms

4.1 LATTICE-BASED CRYPTOGRAPHY

4.1.1 Learning with Errors (LWE)

In the rapidly evolving landscape of cryptography, the emergence of quantum computing poses an existential threat to classical encryption methods. As researchers scramble to develop quantum-resistant algorithms, one approach has gained particular prominence: Learning With Errors (LWE). This mathematical framework, introduced by Oded Regev in 2005, has become a cornerstone of post-quantum cryptography due to its theoretical security guarantees and computational hardness. LWE-based cryptographic schemes offer a promising path forward in an era where traditional public-key cryptography may soon be rendered obsolete by quantum attacks.

Learning with Errors represents a groundbreaking advancement in the field of post-quantum cryptography, offering a robust mathematical foundation for secure communication in the quantum era. Its resistance to quantum attacks, theoretical hardness guarantees, and versatility in cryptographic constructions make it one of the most promising approaches for future security systems. While practical challenges remain in terms of efficiency and implementation, ongoing research and standardization efforts are rapidly addressing these concerns. As the world prepares for the transition to quantum-resistant cryptography, LWE-based schemes are poised to play a central role in safeguarding digital infrastructure against emerging computational threats. The continued development and refinement of these algorithms will be crucial in ensuring a secure and resilient cryptographic landscape for decades to come.

As the field of post-quantum cryptography continues to evolve, research into Learning with Errors and its variants remains an active and vital area of study. One promising direction is the exploration of new mathematical structures that can further enhance the efficiency of LWE-based schemes while maintaining their security properties. This includes investigating alternative error distributions, novel lattice structures, and hybrid approaches that combine LWE with other post-quantum primitives to achieve optimal performance.

Another important research focus is the development of side-channel resistant implementations of LWE-based cryptography. Like all cryptographic systems, LWE-based schemes must be carefully implemented to prevent information leakage through physical side channels, such as power consumption or electromagnetic emissions. Advances in secure implementation techniques will be essential for ensuring that the theoretical security of LWE translates into practical protection against real-world attacks.

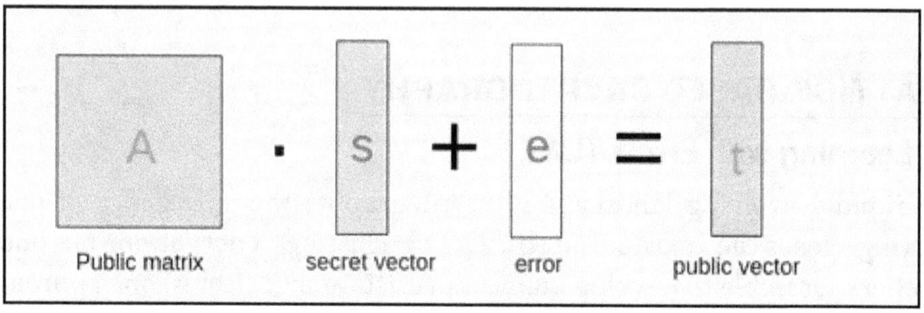

Fig: Learning with Errors: Quantum-Proof Algorithm

Mathematical Foundations

At its core, Learning with Errors is a computational problem that builds upon the concept of learning linear functions in the presence of noise. The problem can be formally described as follows: given a set of linear equations with random coefficients and small random errors, the task is to recover the original linear function. Mathematically, this involves solving for a secret vector s in the equation $y = (a, s) + e$, where a is a random vector, y is the observed output, and e represents a small error term. The introduction of these errors makes the problem computationally hard, even for quantum computers, as it disrupts the linear structure that would otherwise make the problem solvable through standard linear algebra techniques.The hardness of LWE is rooted in the worst-case hardness of lattice problems, particularly the Shortest Vector Problem (SVP) and the Closest Vector Problem (CVP). These lattice problems are believed to be resistant to quantum attacks, making LWE-based cryptography a strong candidate for post-quantum security. The connection between LWE and lattice problems provides a solid theoretical foundation, as breaking LWE would imply solving these fundamental lattice problems, which are conjectured to remain hard even against quantum algorithms like Shor's. This theoretical robustness has made LWE a preferred choice for constructing cryptographic primitives that can withstand future computational threats.

LWE in Cryptographic Constructions

The versatility of the Learning with Errors problem has led to its adoption in various cryptographic constructions, including public-key encryption, digital signatures, and fully homomorphic encryption. One of the most notable applications is in lattice-based cryptography, where LWE serves as the basis for schemes like Regev's encryption and the Ring-LWE variant, which offers improved efficiency. These constructions leverage the hardness of LWE to create cryptographic systems that are both secure and practical for real-world deployment.

In public-key encryption, LWE enables the creation of key pairs where the public key consists of noisy linear equations, and the private key is the secret vector s. The encryption process involves adding additional noise to obscure the message, while decryption uses the private key to remove this noise and recover the original plaintext. This approach ensures that even if an attacker has access to the public key and ciphertexts, the presence of errors makes it computationally infeasible to deduce the private key. Similarly, in digital signatures, LWE-based schemes use the hardness of the problem to prevent forgery, as generating a valid signature requires knowledge of the secret vector that would be difficult to derive from public information alone.

Security Advantages

The primary advantage of Learning With Errors-based cryptography lies in its resistance to quantum attacks. Unlike classical cryptographic schemes such as RSA and ECC, which are vulnerable to Shor's algorithm, LWE-based systems rely on problems that are not known to be efficiently solvable by quantum computers. This resistance stems from the fact that quantum algorithms do not provide a significant speedup for solving the underlying lattice problems that make LWE hard. As a result, LWE-based cryptography is considered one of the most promising approaches for achieving long-term security in the post-quantum era. Another significant benefit of LWE is its flexibility in parameter selection, which allows for trade-offs between security and efficiency.

By adjusting the dimensions of the vectors and the size of the error terms, cryptographers can fine-tune the hardness of the problem to meet specific security requirements. This adaptability makes LWE-based schemes suitable for a wide range of applications, from lightweight cryptography for IoT devices to high-security applications in financial and government sectors. Furthermore, the theoretical foundations of LWE provide confidence in its security, as reductions from worst-case lattice problems ensure that breaking practical instances would require solving computationally intractable problems.

Practical Challenges and Considerations

Despite its theoretical strengths, Learning with Errors-based cryptography faces several practical challenges that must be addressed for widespread adoption. One of the primary concerns is the relatively large key sizes and computational overhead associated with LWE-based schemes. Compared to classical cryptographic algorithms, LWE-based encryption and signatures often require larger keys and more complex operations, which can impact performance and storage requirements. This has led to the development of optimized variants, such as Ring-LWE and Module-LWE, which improve efficiency while maintaining security.

Another challenge is the need for precise parameter selection to balance security and performance. The choice of error distribution, modulus size, and lattice dimensions must be carefully calibrated to ensure that the problem remains hard while allowing for efficient computation. Poorly chosen parameters can either weaken the security of the scheme or make it impractical for real-world use. Ongoing research in this area aims to identify optimal parameter sets that provide strong security guarantees without excessive computational costs.

LWE in Post-Quantum Standardization Efforts

The importance of Learning with Errors in post-quantum cryptography is reflected in its prominent role in standardization efforts. The National Institute of Standards and Technology (NIST) has been leading the charge in evaluating and standardizing post-quantum cryptographic algorithms, with several LWE-based schemes emerging as leading candidates.

Notable examples include Kyber, a key encapsulation mechanism based on Module-LWE, and Dilithium, a digital signature scheme also derived from LWE principles. These algorithms have been selected for their strong security guarantees and practical performance characteristics, making them prime candidates for future cryptographic standards.The inclusion of LWE-based schemes in these standardization efforts underscores their potential to become the foundation of secure communications in the quantum era.

As governments and industries prepare for the transition to post-quantum cryptography, LWE-based solutions are likely to play a central role in securing sensitive data against both classical and quantum threats. The ongoing evaluation and refinement of these algorithms will be crucial in ensuring their readiness for deployment in critical infrastructure and global communication networks.

4.1.2 NTRU

NTRU is an open-source public-key cryptosystem that uses lattice-based cryptography to encrypt and decrypt data. It consists of two algorithms: NTRUEncrypt, which is used for encryption, and NTRUSign, which is used for digital signatures. Unlike other popular public-key cryptosystems, it is resistant to attacks using Shor's algorithm. NTRUEncrypt was patented, but it was placed in the public domain in 2017. NTRUSign is patented, but it can be used by software under the GPL.

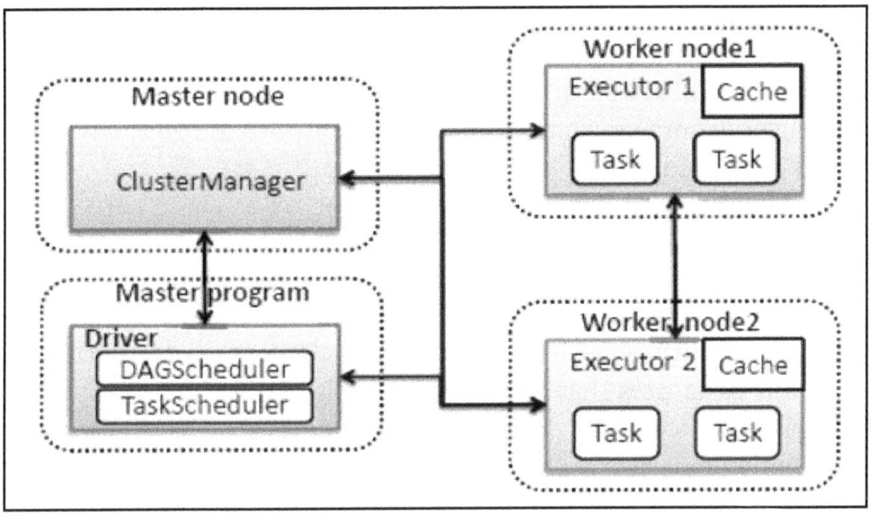

Fig: NTRU Encryption Architecture

History

The first version of the system, which was called NTRU, was developed in 1996 by mathematicians Jeffrey Hoffstein, Jill Pipher, and Joseph H. Silverman. That same year, the developers of NTRU joined with Daniel Lieman and founded the company NTRU Cryptosystems, Inc., and were given a patent on the cryptosystem.

The name "NTRU", chosen for the company and soon applied to the system as well, was originally derived from the pun Number Theorists 'R' Us or, alternatively, stood for Number Theory Research Unit. In 2009, the company was acquired by Security Innovation, a software security corporation. In 2013, Damien Stehle and Ron Steinfeld created a provably secure version of NTRU, which is being studied by a post-quantum crypto group chartered by the European Commission.

In May 2016, Daniel Bernstein, Chitchanok Chuengsatiansup, Tanja Lange and Christine van Vredendaal released NTRU Prime, which adds defenses against a potential attack on NTRU by eliminating algebraic structure they considered worrisome. However, after more than 20 years of scrutiny, no concrete approach to attack the original NTRU by exploiting its algebraic structure has been found so far. NTRU became a finalist in the third round of NIST's Post-Quantum Cryptography Standardization project, whereas NTRU Prime became an alternate candidate.

Performance

At equivalent cryptographic strength, NTRU performs costly private-key operations much faster than RSA does. The time of performing an RSA private operation increases as the cube of the key size, whereas that of an NTRU operation increases quadratically.

In 2010, the Department of Electrical Engineering, University of Leuven, noted that "[using] a modern GTX280 GPU, a throughput of up to 200000 encryptions per second can be reached at a security level of 256 bits. Comparing this to a symmetric cipher (not a very common comparison), this is only around 20 times slower than a recent AES implementation."

Resistance to Quantum-Computer-Based Attacks

Unlike RSA and elliptic-curve cryptography, NTRU is not known to be vulnerable to attacks on quantum computers. The National Institute of Standards and Technology wrote in a 2009 survey that "[there] are viable alternatives for both public key encryption and signatures that are not vulnerable to Shor's Algorithm" and that "[of] the various lattice based cryptographic schemes that have been developed, the NTRU family of cryptographic algorithms appears to be the most practical". The European Union's PQCRYPTO project (Horizon 2020 ICT-645622) is evaluating the provably secure Stehle–Steinfeld version of NTRU (not original NTRU algorithm itself) as a potential European standard. However the Stehle–Steinfeld version of NTRU is "significantly less efficient than the original scheme".

Standardization

- The standard IEEE Std 1363.1, issued in 2008, standardizes lattice-based public-key cryptography, especially NTRUEncrypt.
- The standard X9.98 standardizes lattice-based public-key cryptography, especially NTRUEncrypt, as part of the X9 standards for the financial services industry.
- The PQCRYPTO project of the European Commission is considering standardization of the provably secure Stehle–Steinfeld version of NTRU.

Implementations

Originally, NTRU was only available as a proprietary, for-pay library, and open-source authors were threatened with legal action. It was not until 2011 that the first open-source implementation appeared, and in 2013, Security Innovation exempted open-source projects from having to get a patent license and released an NTRU reference implementation under the GPL v2.

Implementations:

- OpenSSH by default uses NTRU combined with the X25519 ECDH key exchange since August 2022, included in version 9.0.
- The GPL-licensed reference implementation.
- A BSD-licensed library.
- bouncycastle.

- Lokinet was the first onion router implementing NTRU algorithm for its intraweb and End-2-End Encrypted events.
- GoldBug Messenger was the first chat and E-mail client with NTRU algorithm under open-source license, which is based on the Spot-On Encryption Suite Kernels.
- Additionally, wolfSSL provides support for NTRU cipher suites in a lightweight C implementation.

4.1.3 Ring-LWE

In post-quantum cryptography, ring learning with errors (RLWE) is a computational problem which serves as the foundation of new cryptographic algorithms, such as NewHope, designed to protect against cryptanalysis by quantum computers and also to provide the basis for homomorphic encryption. Public-key cryptography relies on construction of mathematical problems that are believed to be hard to solve if no further information is available, but are easy to solve if some information used in the problem construction is known. Some problems of this sort that are currently used in cryptography are at risk of attack if sufficiently large quantum computers can ever be built, so resistant problems are sought.

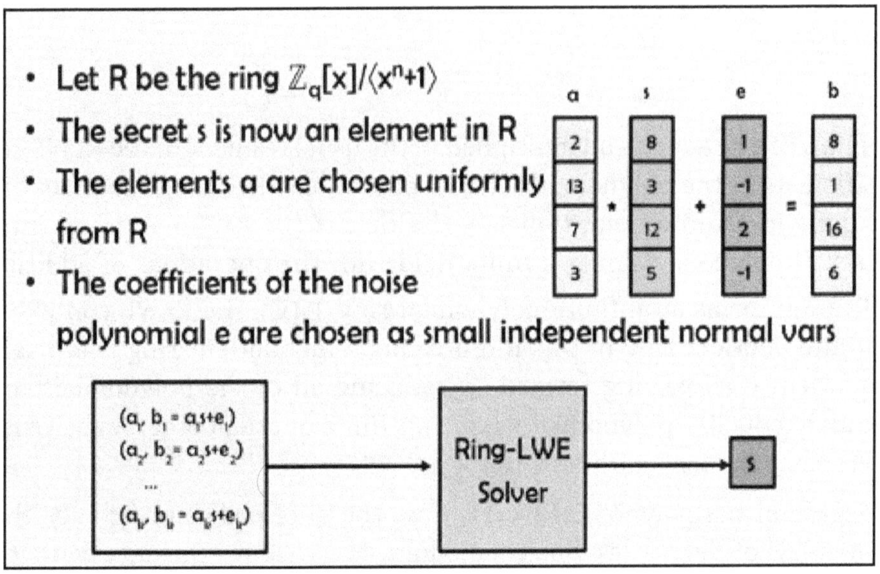

Fig: Ring Learning with Errors

RLWE is more properly called learning with errors over rings and is simply the larger learning with errors (LWE) problem specialized to polynomial rings over finite fields. Because of the presumed difficulty of solving the RLWE problem even on a quantum computer, RLWE based cryptography may form the fundamental base for public-key cryptography in the future just as the integer factorization and discrete logarithm problem have served as the base for public key cryptography since the early 1980s. An important feature of basing cryptography on the ring learning with errors problem is the fact that the solution to the RLWE problem can be used to solve a version of the shortest vector problem (SVP) in a lattice (a polynomial-time reduction from this SVP problem to the RLWE problem has been presented).

Background

The security of modern cryptography, in particular public-key cryptography, is based on the assumed intractability of solving certain computational problems if the size of the problem is large enough and the instance of the problem to be solved is chosen randomly. The classic example that has been used since the 1970s is the integer factorization problem. It is believed that it is computationally intractable to factor the product of two prime numbers if those prime numbers are large enough and chosen at random. As of 2015 research has led to the factorization of the product of two 384-bit primes but not the product of two 512-bit primes. Integer factorization forms the basis of the widely used RSA cryptographic algorithm.

The ring learning with errors (RLWE) problem is built on the arithmetic of polynomials with coefficients from a finite field. A typical polynomial $a(x)$ is expressed as:

$$a(x) = a_0 + a_1 x + a_2 x^2 + \ldots + a_{n-2} x^{n-2} + a_{n-1} x^{n-1}$$

Polynomials can be added and multiplied in the usual fashion. In the RLWE context the coefficients of the polynomials and all operations involving those coefficients will be done in a finite field, typically the field $\mathbf{Z}/q\mathbf{Z} = \mathbf{F}_q$ for a prime integer q q. The set of polynomials over a finite field with the operations of addition and multiplication forms an infinite polynomial ring $(\mathbf{F}_q[x])$. The RLWE context works with a finite quotient ring of this infinite ring. The quotient ring is typically the finite quotient (factor) ring formed by reducing all of the polynomials in $\mathbf{F}_q[x]$ modulo an irreducible polynomial $\Phi(x)$. This finite quotient ring can be written as $\mathbf{F}_q[x]/\Phi(x)$ though many authors write $\mathbf{Z}_q[x]/\Phi(x)$.

If the degree of the polynomial $\Phi(x)$ is n n, the quotient ring becomes the ring of polynomials of degree less than n modulo $\Phi(x)$ with coefficients from F_q. The values n, q, together with the polynomial $\Phi(x)$ partially define the mathematical context for the RLWE problem. Another concept necessary for the RLWE problem is the idea of "small" polynomials with respect to some norm. The typical norm used in the RLWE problem is known as the infinity norm (also called the uniform norm). The infinity norm of a polynomial is simply the largest coefficient of the polynomial when these coefficients are viewed as integers. Hence, $\| a(x) \|_\infty = b$ states that the infinity norm of the polynomial $a(x)$ is b. Thus b is the largest coefficient of $a(x)$.

The final concept necessary to understand the RLWE problem is the generation of random polynomials in $\mathbf{F}_q[x]/\Phi(x)$ and the generation of "small" polynomials . A random polynomial is easily generated by simply randomly sampling the n coefficients of the polynomial from $\mathbf{F}_q$, where $\mathbf{F}_q$ is typically represented as the set $\{-(q-1)/2,...,-1,0,1,...,(q-1)/2\}$.

Randomly generating a "small" polynomial is done by generating the coefficients of the polynomial from $\mathbf{F}_q$ in a way that either guarantees or makes very likely small coefficients. When q is a prime integer, there are two common ways to do this:

- **Using Uniform Sampling:** The coefficients of the small polynomial are uniformly sampled from a set of small coefficients. Let b be an integer that is much less than q. If we randomly choose coefficients from the set: $\{-b,-b+1,-b+2,\ldots,-2,-1,0,1,2,\ldots,b-2,b-1,b\}$ the polynomial will be small with respect to the bound (b).
- **Using Discrete Gaussian Sampling:** For an odd value for q, the coefficients of the polynomial are randomly chosen by sampling from the set $\{-(q-1)/2,\ldots,(q-1)/2\}$ according to a discrete Gaussian distribution with mean 0 0 and distribution parameter σ. The references describe in full detail how this can be accomplished. It is more complicated than uniform sampling but it allows for a proof of security of the algorithm. The paper "Sampling from Discrete Gaussians for Lattice-Based Cryptography on a Constrained Device" by Dwarakanath and Galbraith provides an overview of this problem.

RLWE Cryptography

A major advantage that RLWE based cryptography has over the original learning with errors (LWE) based cryptography is found in the size of the public and private keys. RLWE keys are roughly the square root of keys in LWE. For 128 bits of security an RLWE cryptographic algorithm would use public keys around 7000 bits in length. The corresponding LWE scheme would require public keys of 49 million bits for the same level of security. On the other hand, RLWE keys are larger than the keys sizes for currently used public key algorithms like RSA and Elliptic Curve Diffie-Hellman which require public key sizes of 3072 bits and 256 bits, respectively, to achieve a 128-bit level of security. From a computational standpoint, however, RLWE algorithms have been shown to be the equal of or better than existing public key systems.

Ring Learning with Errors Key Exchanges (RLWE-KEX)

The fundamental idea of using LWE and Ring LWE for key exchange was proposed and filed at the University of Cincinnati in 2011 by Jintai Ding. The basic idea comes from the associativity of matrix multiplications, and the errors are used to provide the security. The paper appeared in 2012 after a provisional patent application was filed in 2012. In 2014, Peikert presented a key transport scheme following the same basic idea of Ding's, where the new idea of sending additional 1 bit signal for rounding in Ding's construction is also utilized. An RLWE version of the classic MQV variant of a Diffie-Hellman key exchange was later published by Zhang et al. The security of both key exchanges is directly related to the problem of finding approximate short vectors in an ideal lattice.

Ring Learning with Errors Signature (RLWE-SIG)

A RLWE version of the classic Feige–Fiat–Shamir Identification protocol was created and converted to a digital signature in 2011 by Lyubashevsky. The details of this signature were extended in 2012 by Gunesyu, Lyubashevsky, and Popplemann in 2012 and published in their paper "Practical Lattice Based Cryptography – A Signature Scheme for Embedded Systems." These papers laid the groundwork for a variety of recent signature algorithms some based directly on the ring learning with errors problem and some which are not tied to the same hard RLWE problems.

Ring Learning with Errors Homomorphic Encryption (RLWE-HOM)

Homomorphic encryption is type of encryption that allows computations to be performed on encrypted data without first having to decrypt it. The purpose of homomorphic encryption is to allow the computations on sensitive data to occur on computing devices that should not be trusted with the data. These computing devices are allowed to process the ciphertext which is output from a homomorphic encryption. In 2011, Brakersky and Vaikuntanathan, published "Fully Homomorphic Encryption from Ring-LWE and Security for Key Dependent Messages" which builds a homomorphic encryption scheme directly on the RLWE problem.

4.1.4 Kyber

Kyber is a key encapsulation mechanism (KEM) designed to be resistant to cryptanalytic attacks with future powerful quantum computers. It is used to establish a shared secret between two communicating parties without an (IND-CCA2) attacker in the transmission system being able to decrypt it. This asymmetric cryptosystem uses a variant of the learning with errors lattice problem as its basic trapdoor function. It won the NIST competition for the first post-quantum cryptography (PQ) standard. NIST calls its standard, numbered FIPS 203, Module-Lattice-Based Key-Encapsulation Mechanism (ML-KEM).

Properties

The system is based on the module learning with errors (M-LWE) problem, in conjunction with cyclotomic rings. Recently, there has also been a tight formal mathematical security reduction of the ring-LWE problem to MLWE. Compared to competing PQ methods, it has typical advantages of lattice-based methods, e.g. in regard to runtime as well as the size of the ciphertexts and the key material.

Variants with different security levels have been defined: Kyber512 (NIST security level 1, ≈AES 128), Kyber768 (NIST security level 3, ≈AES 192), and Kyber1024 (NIST security level 5, ≈AES 256). At the Kyber768 level, the secret keys are 2400 bytes in size, the public keys 1184, and the ciphertexts 1088.

With an accordingly optimized implementation, 4 kilobytes of memory can be sufficient for the cryptographic operations. For a chat encryption scenario using liboqs, replacing the extremely efficient, non-quantum-safe ECDH key exchange using Curve25519 was found to increase runtime by a factor of about 2.3 (1.5–7), an estimated 2.3-fold (1.4–3.1) increase in energy consumption, and have about 70 times (48–92) more data overhead. Internal hashing operations account for the majority of the runtime, which would thus potentially benefit greatly from corresponding hardware acceleration.

Development

Kyber is derived from a method published in 2005 by Oded Regev, developed by developers from Europe and North America, who are employed by various government universities or research institutions, or by private companies, with funding from the European Commission, Switzerland, the Netherlands, and Germany. They also developed the related and complementary signature scheme Dilithium, as another component of their "Cryptographic Suite for Algebraic Lattices" (CRYSTALS).

Like other PQC-KEM methods, Kyber makes extensive use of hashing internally. In Kyber's case, variants of Keccak (SHA-3/SHAKE) are used here, to generate pseudorandom numbers, among other things. In 2017 the method was submitted to the US National Institute of Standards and Technology (NIST) for its public selection process for a first standard for quantum-safe cryptographic primitives (NISTPQC). It is the only key encapsulation mechanism that has been selected for standardization at the end of the third round of the NIST standardization process.

According to a footnote the report announcing the decision, it is conditional on the execution of various patent-related agreements, with NTRU being a fallback option. Currently, a fourth round of the standardization process is underway, with the goal of standardizing an additional KEM. In the second phase of the selection process, several parameters of the algorithm were adjusted and the compression of the public keys was dropped. Most recently, NIST paid particular attention to costs in terms of runtime and complexity for implementations that mask runtimes in order to prevent corresponding side-channel attacks (SCA).

Evolution

Kyber underwent changes during the NIST standardization process. In particular, in the submission for round 2 (so called Kyber v2), the following features were changed:

- Public key compression removed (due to NIST comments on the security proof).
- Parameter q reduced to 3329 (from 7681).
- Ciphertext compression parameters changed.
- Number-theoretic transform (NTT) definition changed along the lines of NTTRU for faster polynomial multiplication,
- Noise parameter reduced to $\eta = 2$ for faster noise sampling.
- Public key representation changed to NTT domain in order to save the NTT operations.

Submission to round 3 underwent further tweaks:

- The use of Fujisaki–Okamoto transformation (FO transform) modified.
- Noise level increased and ciphertext compression reduced for the level 1 parameter set.
- Sampling algorithm improved.

Usage

The developers have released a reference implementation into the public domain (or under CC0), which is written in C. The program library liboqs of the Open Quantum Safe (OQS) project contains an implementation based on that. OQS also maintains a quantum-safe Provider module for OpenSSL 3.x, and has integrated its code into BoringSSL and wolfSSL. There are a handful of implementations using various other programming languages from third-party developers, including JavaScript and Java. Various (free) optimized hardware implementations exist, including one that is resistant to side-channel attacks.

The German Federal Office for Information Security is aiming for implementation in Thunderbird, and in this context also an implementation in the Botan program library and corresponding adjustments to the OpenPGP standard. Amazon Web Services (AWS) integrated Kyber into their Key Management Service (KMS) in 2020 as a hybrid post-quantum key exchange option for TLS connections. In 2023, the encrypted messaging service Signal implemented PQXDH, a Kyber-based post-quantum encryption algorithm, to their Signal Protocol.

4.1.5 Dilithium

Dilithium, Li_2, is a strongly electrophilic, diatomic molecule comprising two lithium atoms covalently bonded together. Li_2 has been observed in the gas phase. It has a bond order of 1, an internuclear separation of 267.3 pm and a bond energy of 102 kJ/mol or 1.06 eV in each bond. The electron configuration of Li_2 may be written as σ^2. Being the third-lightest stable neutral homonuclear diatomic molecule (after dihydrogen and dihelium), dilithium is an extremely important model system for studying fundamentals of physics, chemistry, and electronic structure theory.

It is the most thoroughly characterized compound in terms of the accuracy and completeness of the empirical potential energy curves of its electronic states. Analytic empirical potential energy curves have been constructed for the X-state, a-state, A-state, c-state, B-state, 2d-state, l-state, E-state, and the F-state. The most reliable of these potential energy curves are of the Morse/Long-range variety. Li_2 potentials are often used to extract atomic properties. For example, the C_3 value for atomic lithium extracted from the A-state potential of Li_2 by Le Roy et al. in is more precise than any previously measured atomic oscillator strength. This lithium oscillator strength is related to the radiative lifetime of atomic lithium and is used as a benchmark for atomic clocks and measurements of fundamental constants.

Electronic state	Spectroscopic symbol	Term symbol	Bond length (pm)		Dissociation energy (cm^{-1})		Bound vibrational levels
1 (Ground)	X	$1^1\Sigma_g^+$	267	.298 74(19)	8 516	.780 0(23)	39
2	a	$1^3\Sigma_u^+$	417	.000 6(32)	333	.779 5(62)	11
3	b	$1^3\Pi_u$					
4	A	$1^1\Sigma_g^+$	310	.792 88(36)	9 353	.179 5 (28)	118
5	c	$1^3\Sigma_g^+$	306	.543 6(16)	7 093	.492 6(86)	104
6	B	$1^1\Pi_u$	293	.617 142(310)	2 984	.444	118
7	E	$3(?)^1\Sigma_g^+$					

4.1.6 Falcon (NIST Finalists)

Falcon is a post-quantum signature scheme selected by the NIST at the fourth round of the post-quantum standardisation process. It was designed by Thomas Prest, Pierre-Alain Fouque, Jeffrey Hoffstein, Paul Kirchner, Vadim Lyubashevsky, Thomas Pornin, Thomas Ricosset, Gregor Seiler, William Whyte, and Zhenfei Zhang. It relies on the hash-and-sign technique over the Gentry, Peikert, and Vaikuntanathan framework over NTRU lattices. The name Falcon is an acronym for Fast Fourier lattice-based compact signatures over NTRU.

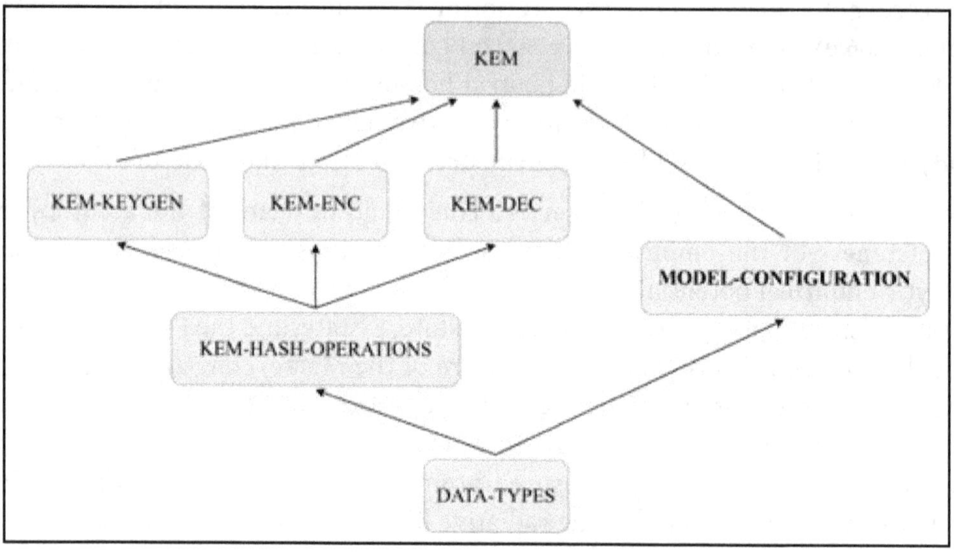

Fig: Falcon Post-Quantum Digital Signature Operation Model

Properties

The design rationale of Falcon takes advantage of multiple tools to ensure compactness and efficiency with provable security. To achieve this goal, the use of a NTRU lattice allows the size of the signatures and public-key to be relatively small, while fast Fourier sampling permits efficient signature computations. From a security point of view, the Gentry, Peikert, and Vaikuntanathan framework enjoys a security reduction in the Quantum Random Oracle Model.

Implementations

The authors of Falcon provide a reference implementation in C as required by the NIST and one in Python for simplicity. The set of parameters suggested by Falcon imply a signature size of 666 bytes and a public key size of 897 bytes for the NIST security level 1 (security comparable to breaking AES-128 bits). The key generation can be performed in 8.64 ms with a throughput of approximately 6,000 signature per second and 28,000 verifications per second. On the other hand, the NIST security level 5 (comparable to breaking AES-256) requires a signature size of 1,280 bytes and a public key size of 1793 bytes, a key generation under 28 ms, and a throughput of 2,900 signatures per second and 13,650 verifications per second.

4.2 CODE-BASED CRYPTOGRAPHY

4.2.1 McEliece and Its Variants

In the evolving landscape of post-quantum cryptography, the McEliece cryptosystem stands as one of the oldest yet most resilient encryption schemes resistant to quantum attacks. Introduced by Robert McEliece in 1978, this code-based cryptographic system has endured decades of cryptanalysis while maintaining its security guarantees. Unlike many classical encryption methods vulnerable to Shor's algorithm, McEliece and its variants rely on the hardness of decoding random linear codes, a problem believed to remain intractable even for quantum computers. As the cryptographic community seeks quantum-resistant alternatives, the McEliece cryptosystem and its modern adaptations have gained renewed attention for their theoretical robustness and practical applications.

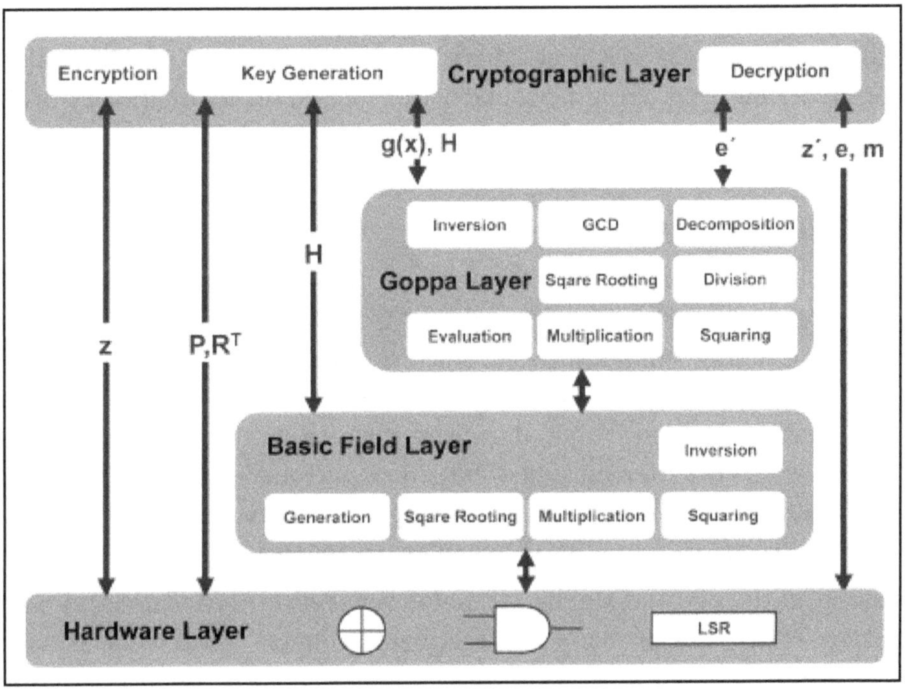

Fig: McEliece Layer Model

The McEliece cryptosystem and its variants represent a cornerstone of code-based cryptography, offering robust security guarantees rooted in well-established coding theory problems. While the original scheme faces practical limitations in terms of key sizes and computational efficiency, modern adaptations have made significant strides in addressing these challenges. As the cryptographic community prepares for the post-quantum era, code-based approaches provide a compelling option for securing communications against future computational threats.

The continued development and standardization of these schemes will be crucial in shaping the cryptographic landscape of tomorrow. With ongoing research and implementation improvements, McEliece and its variants are poised to play an important role in the next generation of secure communication systems. The field of code-based cryptography continues to evolve, with ongoing research exploring new directions to improve both security and efficiency. One promising area of investigation involves the development of new code families that could offer better performance characteristics while maintaining strong security guarantees. Researchers are also examining hybrid approaches that combine code-based cryptography with other post-quantum primitives to achieve optimal security and efficiency trade-offs.

Another important research focus is the development of more efficient decoding algorithms that could be used in the decryption process without compromising security. Advances in this area could significantly improve the practicality of code-based schemes for resource-constrained environments. Additionally, the exploration of new cryptographic constructions based on coding theory problems may lead to innovative applications beyond traditional encryption and signature schemes.

Original McEliece Cryptosystem

The McEliece cryptosystem represents a pioneering approach to public-key encryption based on error-correcting codes. At its core, the system leverages the difficulty of decoding random linear codes, specifically Goppa codes, to achieve security. The encryption process begins with the generation of a generator matrix G for a Goppa code, which is then disguised through matrix operations to create the public key. To encrypt a message, the sender introduces controlled errors to a codeword derived from the public key, creating a ciphertext that only the legitimate recipient can decode using the private key's error-correcting capabilities.

The security of the original McEliece scheme relies on two fundamental hardness assumptions. First, the problem of decoding a general linear code is NP-hard, meaning no efficient algorithm is known to solve it, even with quantum computers. Second, the specific structure of Goppa codes remains hidden through the public key's transformation, preventing attackers from exploiting any algebraic properties. This combination of computational hardness and structural obfuscation has allowed the McEliece cryptosystem to withstand decades of cryptanalysis, maintaining its status as a secure encryption method despite its age.

Security Foundations and Cryptanalysis

The enduring security of the McEliece cryptosystem stems from its foundation in coding theory problems that have resisted efficient solutions for decades. The primary security assumption is that decoding a random linear code is computationally infeasible, a problem that remains hard even when considering quantum algorithms. Unlike lattice-based or hash-based cryptographic schemes that rely on relatively newer mathematical constructs, code-based cryptography benefits from well-established theoretical foundations in information theory and error correction.

Over the years, various cryptanalytic attacks have been proposed against the McEliece cryptosystem, each addressing different aspects of its structure. Information-set decoding attacks attempt to exploit the error-correcting properties of the code, while structural attacks target the specific properties of Goppa codes. However, these attacks generally require exponential time complexity, making them impractical for real-world implementations with properly chosen parameters. The resilience of McEliece against these attacks has contributed to its longevity as a secure cryptographic primitive, particularly as other post-quantum candidates face more substantial cryptanalytic progress.

Practical Challenges and Limitations

Despite its strong security foundations, the original McEliece cryptosystem faces several practical challenges that have limited its widespread adoption. The most significant obstacle is the large key sizes required to achieve adequate security levels. Public keys in McEliece encryption typically range from hundreds of kilobytes to several megabytes, making them impractical for many modern applications where bandwidth and storage are constrained. This key size issue stems from the need to maintain security against information-set decoding attacks, which become more feasible as key sizes decrease.

Another practical limitation is the computational overhead associated with encryption and decryption operations. The matrix operations required for both processes can be resource-intensive, particularly for resource-constrained devices such as IoT sensors or mobile applications. These performance characteristics have historically made McEliece less attractive compared to more efficient cryptographic schemes, despite its strong security guarantees. The trade-off between security and efficiency has been a persistent challenge for code-based cryptography in general, requiring careful parameter selection to balance these competing requirements.

Modern Variants and Improvements

Recognizing the potential of code-based cryptography while addressing its practical limitations, researchers have developed several variants of the McEliece cryptosystem that improve efficiency while maintaining security. One notable variant is the Niederreiter encryption scheme, which presents a dual formulation of McEliece using parity-check matrices instead of generator matrices. This approach offers equivalent security with potentially more efficient implementations, particularly in the encryption process.

More recent developments have focused on using different families of error-correcting codes to reduce key sizes while preserving security. Quasi-cyclic (QC) codes and their extensions have shown particular promise in this regard. By introducing algebraic structure through quasi-cyclic properties, these variants achieve significant key size reductions compared to the original McEliece scheme. The QC-MDPC (Quasi-Cyclic Moderate Density Parity Check) variant, for instance, has demonstrated practical key sizes while maintaining resistance to known attacks. These modern adaptations represent important steps toward making code-based cryptography more viable for real-world applications.

Standardization Efforts and Post-Quantum Considerations

The renewed interest in code-based cryptography has led to the inclusion of McEliece variants in post-quantum standardization efforts. The National Institute of Standards and Technology (NIST) has evaluated several code-based cryptographic schemes as part of its Post-Quantum Cryptography Standardization Project. While the original McEliece cryptosystem was not selected as a finalist due to its large key sizes, its modern variants have shown more promise.

The Classic McEliece scheme, which maintains the original structure but with optimized parameters, was selected as an alternate candidate in the NIST standardization process, recognizing its strong security foundations. The inclusion of code-based schemes in these standardization efforts highlights their potential role in future cryptographic infrastructures. As quantum computing advances, the need for quantum-resistant algorithms becomes increasingly urgent. Code-based cryptography, with its long history of security and theoretical robustness, offers a compelling option for securing communications in the post-quantum era. The ongoing research and development of these schemes will be crucial in determining their ultimate place in the cryptographic landscape.

Implementation Considerations

Implementing McEliece and its variants in practical systems requires careful attention to several factors. The choice of parameters, particularly the code length and error weight, significantly impacts both security and performance. Larger parameters generally provide stronger security but at the cost of increased key sizes and computational requirements. Finding the optimal balance between these factors is essential for creating efficient yet secure implementations. Side-channel resistance is another critical consideration in the implementation of code-based cryptographic schemes. Like all cryptographic algorithms, McEliece variants must be implemented in ways that prevent information leakage through physical observations. Constant-time implementations and other countermeasures are necessary to protect against timing attacks and other side-channel exploits. The development of secure implementation practices will be vital for the real-world deployment of these cryptographic primitives.

Comparative Analysis with Other Post-Quantum Approaches

When evaluating McEliece and its variants against other post-quantum cryptographic approaches, several comparative advantages and disadvantages emerge. Compared to lattice-based cryptography, code-based schemes generally offer stronger theoretical security guarantees based on well-established coding theory problems. However, lattice-based approaches often provide better performance characteristics and smaller key sizes, making them more practical for many applications. Hash-based cryptography, another post-quantum candidate, typically offers excellent performance but may lack the rich functionality provided by code-based schemes. The choice between these approaches often depends on the specific requirements of the application, with code-based cryptography offering a compelling option when strong theoretical security is prioritized over implementation efficiency. This comparative analysis underscores the importance of having diverse cryptographic primitives available to address different security needs in the post-quantum landscape.

4.3 MULTIVARIATE POLYNOMIAL CRYPTOGRAPHY

4.3.1 Rainbow

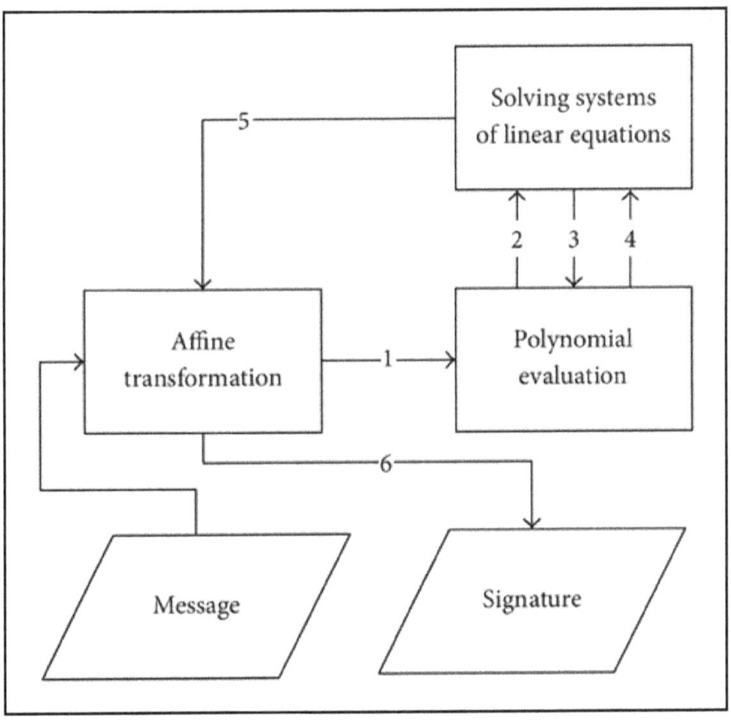

Fig: The Flowchart of Implementations of Rainbow Scheme

In the quest for quantum-resistant cryptographic solutions, multivariate polynomial cryptography has emerged as a promising approach that leverages the complexity of solving systems of nonlinear equations. Among the various schemes in this category, Rainbow stands out as a particularly sophisticated and well-studied digital signature algorithm. Developed as an evolution of earlier multivariate cryptographic systems, Rainbow combines mathematical elegance with practical security considerations, making it a strong candidate for post-quantum cryptography. As classical cryptographic systems face existential threats from quantum computing, understanding multivariate approaches like Rainbow becomes increasingly important for developing secure digital infrastructures.

The Rainbow signature scheme represents a significant advancement in multivariate polynomial cryptography, offering a sophisticated approach to post-quantum digital signatures. Its layered construction improves upon earlier multivariate schemes by providing better performance characteristics while maintaining strong security guarantees. As one of the most well-studied multivariate cryptographic systems, Rainbow demonstrates the potential of this approach to contribute to the future of secure digital communications. While challenges remain in terms of implementation efficiency and parameter selection, ongoing research continues to address these issues. The mathematical foundations of multivariate cryptography, combined with the practical improvements embodied in Rainbow, position this approach as an important component of the post-quantum cryptographic landscape. As the field evolves, schemes like Rainbow will play a crucial role in developing the quantum-resistant cryptographic systems of tomorrow.

The field of multivariate polynomial cryptography continues to evolve, with Rainbow serving as an important milestone in this development. Future research directions include exploring new polynomial structures that could offer improved security or efficiency characteristics. Investigations into alternative trapdoor constructions may lead to schemes with better performance profiles while maintaining strong security guarantees. Additionally, the study of new cryptanalytic techniques will be crucial for understanding the long-term security of these schemes and guiding parameter selection.

Another promising area of research involves hybrid approaches that combine multivariate cryptography with other post-quantum primitives. These combinations could potentially leverage the strengths of multiple cryptographic approaches to achieve optimal security and efficiency trade-offs. The development of more efficient algorithms for polynomial operations could also significantly improve the practicality of multivariate schemes. As quantum computing advances, the continued evolution of multivariate cryptography will be essential for maintaining secure digital communications in the post-quantum era.

Mathematical Foundations

Multivariate polynomial cryptography operates on the principle that solving systems of nonlinear polynomial equations is computationally hard, particularly when the equations are constructed in a specific manner. The security of these systems relies on the difficulty of solving the Multivariate Quadratic (MQ) problem, which involves finding solutions to a set of quadratic polynomial equations over a finite field. This problem is known to be NP-hard in the general case, meaning no efficient algorithm is known to solve it, even with quantum computers. The Rainbow signature scheme builds upon this foundation by carefully constructing polynomial systems that are easy to evaluate but hard to invert, creating an asymmetric structure suitable for digital signatures.

The mathematical structure of Rainbow involves multiple layers of quadratic polynomials, where each layer consists of polynomials that map from one finite field to another. This layered construction allows for efficient signature generation while maintaining the hardness of the underlying MQ problem. The polynomials are designed to have a trapdoor structure that enables the legitimate signer to compute signatures efficiently, while an attacker without knowledge of the trapdoor faces the computationally intractable task of solving the polynomial system. This balance between efficiency and security forms the core of Rainbow's cryptographic strength.

Rainbow Signature Scheme

Rainbow represents a significant advancement in multivariate cryptography by introducing a layered construction that improves both security and efficiency compared to earlier schemes. The signature generation process involves evaluating a sequence of quadratic polynomial maps, where each layer reduces the dimensionality of the problem. This construction allows for shorter signatures and faster verification compared to previous multivariate signature schemes. The security of Rainbow relies on the difficulty of solving the underlying system of quadratic equations, which is believed to be resistant to both classical and quantum attacks.

The scheme operates by first generating a set of public polynomials that form the public key, while the private key consists of the trapdoor information that allows for efficient inversion of these polynomials. To sign a message, the signer uses the private key to solve a modified version of the public polynomial system, producing a signature that can be verified using the public key. The layered structure of Rainbow enables this process to be more efficient than earlier multivariate schemes while maintaining strong security guarantees. This combination of practical performance and theoretical security has made Rainbow a leading candidate in the field of post-quantum digital signatures.

Security Analysis and Cryptographic Strength

The security of Rainbow is based on several interconnected hardness assumptions that contribute to its resistance against known cryptanalytic attacks. The primary security assumption is the hardness of solving the underlying system of multivariate quadratic equations, which forms the basis of the MQ problem. This problem has been extensively studied in computational algebra and is known to be NP-hard, providing a strong theoretical foundation for the scheme's security. Additionally, the specific structure of Rainbow's polynomial layers introduces further complexity that enhances security beyond that of generic MQ instances.

Cryptanalytic research on Rainbow has focused on various attack vectors, including direct algebraic attacks, min-rank attacks, and differential attacks. These analyses have generally confirmed the scheme's resistance to known attacks when properly parameterized. The layered construction of Rainbow provides additional security by making it more difficult for attackers to exploit any single layer's structure. However, like all cryptographic schemes, Rainbow requires careful parameter selection to balance security and efficiency. The ongoing study of these parameters and potential attack vectors remains an active area of research in the multivariate cryptography community.

Performance Characteristics and Practical Considerations

One of the most significant advantages of Rainbow over earlier multivariate signature schemes is its improved performance characteristics. The layered construction allows for shorter signature sizes and faster verification times, addressing two of the primary practical limitations of multivariate cryptography.

Signature sizes in Rainbow are typically smaller than those in previous schemes like Unbalanced Oil and Vinegar (UOV), making them more suitable for real-world applications where bandwidth and storage are concerns. The computational efficiency of Rainbow is particularly notable in the verification process, which involves evaluating a sequence of quadratic polynomials. This operation is significantly faster than solving the polynomial system required for signature generation, making Rainbow well-suited for scenarios where verification efficiency is critical. However, signature generation remains more computationally intensive than in some other digital signature schemes, which may limit its applicability in resource-constrained environments. The trade-off between signature size, verification speed, and signature generation time represents an important consideration in the practical deployment of Rainbow.

Parameter Selection and Security Trade-offs

The security and efficiency of Rainbow are heavily dependent on the careful selection of its parameters, which determine the structure and complexity of the underlying polynomial system. These parameters include the number of layers, the dimensions of the polynomial maps, and the characteristics of the finite fields used. The choice of these parameters involves complex trade-offs between security, signature size, and computational efficiency. Larger parameters generally provide stronger security but result in larger signatures and slower operations, while smaller parameters improve performance at the potential cost of reduced security margins.

The process of parameter selection for Rainbow involves both theoretical analysis and practical experimentation to ensure resistance against known attacks while maintaining acceptable performance. Cryptographers must consider the latest developments in cryptanalysis when choosing parameters, as new attack techniques may require adjustments to maintain security. This dynamic relationship between parameter selection and security analysis highlights the importance of ongoing research in multivariate cryptography. The Rainbow scheme's flexibility in parameter choices allows it to be adapted to different security requirements, making it a versatile option for various applications.

Comparison with Other Post-Quantum Signature Schemes

When evaluating Rainbow in the context of other post-quantum digital signature schemes, several comparative advantages and disadvantages become apparent. Compared to hash-based signature schemes, Rainbow offers potentially stronger security guarantees based on well-studied mathematical problems rather than hash function properties. However, hash-based schemes often provide better performance characteristics, particularly in terms of signature generation speed. Lattice-based signature schemes, another prominent post-quantum candidate, typically offer a different balance of security and efficiency, with some schemes providing smaller signatures but potentially different security assumptions.

The choice between these approaches often depends on the specific requirements of the application. Rainbow's strength lies in its mathematical foundations and the extensive cryptanalysis it has undergone, providing confidence in its security properties. For applications where signature size and verification efficiency are critical, Rainbow presents a compelling option among post-quantum signature schemes. The diversity of post-quantum approaches ensures that different security needs can be addressed, with multivariate schemes like Rainbow playing an important role in this cryptographic ecosystem.

Implementation Challenges and Considerations

Implementing Rainbow in practical systems requires careful attention to several factors to ensure both security and performance. The complexity of the polynomial operations involved demands efficient algorithms for both signature generation and verification. Optimized implementations must balance computational efficiency with resistance to side-channel attacks, which could potentially leak information about the private key. Constant-time implementations and other countermeasures are essential to protect against timing attacks and other physical observations that might compromise security. Another implementation consideration is the management of the relatively large public keys characteristic of multivariate schemes. While Rainbow has made significant improvements in this area compared to earlier schemes, key management remains an important practical concern. The storage and transmission of these keys must be handled efficiently, particularly in resource-constrained environments. These implementation challenges highlight the need for continued research into optimization techniques and secure implementation practices for multivariate cryptographic schemes.

Standardization Efforts and Cryptographic Evaluation

Rainbow has been recognized as a strong candidate in the ongoing efforts to standardize post-quantum cryptographic algorithms. The National Institute of Standards and Technology (NIST) included Rainbow in its Post-Quantum Cryptography Standardization Project, subjecting it to rigorous evaluation alongside other promising schemes. This evaluation process involves extensive cryptanalysis, performance testing, and implementation scrutiny to assess the scheme's suitability for standardization. While Rainbow was not selected as a finalist in the most recent round of NIST's standardization process, the evaluation provided valuable insights into its strengths and potential areas for improvement.

The cryptographic community continues to study and refine multivariate signature schemes like Rainbow, recognizing their potential contributions to post-quantum security. Ongoing research focuses on addressing the practical limitations of these schemes while maintaining their strong security properties. The lessons learned from the standardization process will inform future developments in multivariate cryptography, potentially leading to new variants or improved implementations that could play a role in future cryptographic standards.

4.3.2 Unbalanced Oil

In the search for quantum-resistant cryptographic schemes, multivariate public key cryptography (MPKC) has emerged as a prominent candidate. MPKC is based on the mathematical hardness of solving systems of nonlinear multivariate polynomial equations over finite fields, a problem known to be NP-hard.

Unlike traditional number-theoretic approaches such as RSA and ECC, MPKC offers an entirely different foundation that is believed to be resistant to quantum attacks. One of the most influential families of MPKC schemes is the Oil and Vinegar (OV) signature scheme, introduced by Jacques Patarin in 1997. The OV scheme and its subsequent refinements, especially the Unbalanced Oil and Vinegar (UOV) scheme, are particularly important because they provide a solid framework for constructing secure digital signatures that are both efficient and post-quantum secure.

$$
\begin{aligned}
y_1 &= f_1{}^*(x_1, x_2, \ldots, x_n) \\
y_2 &= f_2{}^*(x_1, x_2, \ldots, x_n) \\
&\vdots \\
y_m &= f_m{}^*(x_1, x_2, \ldots, x_n)
\end{aligned}
$$

Fig: Unbalanced Oil and Vinegar Scheme

Unbalanced Oil and Vinegar represents a powerful and resilient approach to constructing digital signatures in a post-quantum world. Its reliance on the intractability of solving multivariate quadratic systems provides a distinct and robust foundation, separate from number-theoretic cryptography. While it faces challenges in terms of key size and implementation, its efficiency and quantum resistance make it a compelling candidate for secure applications. As research progresses and standardization efforts continue, UOV is likely to remain a cornerstone of multivariate cryptographic research and a valuable component of future cryptographic systems prepared to face the quantum threat.

The UOV scheme holds a significant place in the theoretical landscape of cryptography. It exemplifies how algebraic hardness assumptions can be translated into practical cryptographic tools. The study of UOV has advanced the understanding of multivariate cryptography and inspired a broad range of related research, including identification protocols, zero-knowledge proofs, and advanced signature schemes. Its enduring relevance, despite the emergence of other quantum-safe candidates, speaks to the robustness of its foundational design. In the coming years, continued improvements in algebraic methods, optimization techniques, and secure implementation practices are expected to make UOV and its variants even more competitive in the post-quantum era.

Oil and Vinegar Construction

The original Oil and Vinegar scheme was developed to provide a trapdoor function based on the difficulty of solving systems of multivariate quadratic equations. The scheme involves two types of variables: "oil" and "vinegar". Vinegar variables are chosen first, and their values are substituted into the equations. Then, because the structure of the equations ensures that they become linear in the oil variables, the oil variables can be solved efficiently. The public key consists of the full system of quadratic equations, while the private key includes the transformation that relates this system to a simpler, structured version. This hidden structure acts as the trapdoor that allows the key holder to solve the otherwise hard problem of inverting the system. The OV scheme was attractive due to its simplicity and reliance on algebraic hardness assumptions, but early versions were found to be vulnerable to attacks exploiting the symmetry and structure of the system.

Emergence of Unbalanced Oil and Vinegar

To address the vulnerabilities in the original OV construction, Patarin introduced the Unbalanced Oil and Vinegar (UOV) scheme. The central idea behind UOV is to increase the number of vinegar variables relative to oil variables. This asymmetry removes much of the exploitable structure that made the original OV scheme vulnerable. By having more vinegar variables than oil variables, the attacker's task becomes significantly more complex, as the increased randomness introduced by the vinegar variables makes it difficult to find relationships between the equations. The UOV scheme maintains the trapdoor functionality while offering greater resistance to known cryptanalytic methods. Its simplicity, efficiency, and apparent resilience to both classical and quantum attacks have made it one of the leading candidates in post-quantum signature schemes.

Structure and Security of UOV Schemes

The UOV scheme functions by defining a system of multivariate quadratic polynomials over a finite field. The signer possesses a private key that includes two affine transformations and a secret central map with a known structure. This central map is designed so that, when vinegar values are fixed, the resulting system becomes linear in oil variables and can be solved efficiently. The public key is derived by composing the central map with the affine transformations, resulting in a seemingly random system of quadratic equations.

Security lies in the difficulty of inverting this system without the trapdoor. The best known attacks against UOV are direct algebraic attacks and rank-based attacks, but with appropriate parameter choices, the UOV scheme has withstood decades of scrutiny. Importantly, no efficient quantum algorithm is known that can solve the general multivariate quadratic problem faster than classical methods, making UOV a strong candidate for quantum-resistant cryptography.

Applications and Implementation of UOV Signatures

UOV is primarily used for digital signatures rather than encryption due to its structure. The signing process is fast and lightweight, involving solving a small system of linear equations. Verification, which involves evaluating quadratic polynomials, is also efficient.

These features make UOV suitable for applications requiring high-speed signature generation and verification, such as embedded systems, IoT devices, and blockchain technologies. UOV signatures have been implemented in various cryptographic libraries and tested for their performance across different hardware platforms. One of the practical advantages of UOV is that the signing process can be parallelized, making it scalable for high-throughput systems. Moreover, the scheme can be tailored to different security levels by adjusting the number of oil and vinegar variables, allowing for a flexible trade-off between performance and security.

Drawbacks and Challenges of UOV

Despite its advantages, UOV also faces several challenges. One of the primary drawbacks is the large size of its public key, which can be several hundred kilobytes in length. This makes storage and transmission inefficient compared to traditional cryptographic schemes. Efforts have been made to compress the public key and reduce its size using techniques such as structured affine maps and isomorphisms of polynomials, but these often introduce new security risks by making the system more predictable. Another concern is the potential for side-channel attacks, especially in constrained devices where implementations may leak information through timing or power analysis. As with all cryptographic schemes, careful implementation is critical to maintaining security. The need to balance compactness, efficiency, and security continues to drive research into optimizing UOV for practical use.

Standardization and the Role of UOV in Post-Quantum Cryptography

UOV is one of the signature schemes submitted to the NIST Post-Quantum Cryptography Standardization Project. While lattice-based schemes like CRYSTALS-Dilithium and hash-based schemes like SPHINCS+ have gained more attention in the initial stages, UOV remains under evaluation and is considered a viable alternative due to its different underlying hardness assumptions. The diversity of assumptions is critical in cryptographic ecosystems, as it helps to ensure that a single breakthrough does not compromise all standardized schemes. UOV's inclusion helps promote a robust and diversified post-quantum cryptographic infrastructure. Even if not selected in the final round, UOV and its variants provide valuable fallback options and contribute to the collective understanding of quantum-resistant signature design.

Recent Advances and Variants of UOV

Several improvements and extensions to UOV have been proposed to address its limitations and enhance its practicality. One such variant is Rainbow, a multivariate signature scheme based on layered UOV constructions. Rainbow improves upon UOV by introducing a hierarchical structure of variables, which reduces key sizes and increases signing efficiency. However, Rainbow has also faced cryptanalytic challenges, including successful key recovery attacks under certain parameter settings, highlighting the delicate balance between efficiency and security. Another approach involves exploring UOV constructions over different finite fields or with different algebraic structures to optimize performance and resilience. Research continues into new variants that preserve the foundational security of UOV while improving its applicability in constrained environments.

4.3.3 Vinegar (UOV)

Multivariate Public Key Cryptography (MPKC) represents a class of post-quantum cryptographic schemes based on the difficulty of solving systems of multivariate polynomial equations over finite fields. Among the various MPKC constructions, the Unbalanced Oil and Vinegar (UOV) signature scheme, commonly referred to simply as "Vinegar," has emerged as a promising candidate for quantum-resistant digital signatures. First proposed by Jacques Patarin in 1997, UOV builds upon the original Oil and Vinegar scheme while addressing its security weaknesses through an unbalanced variable structure. The system derives its name from the metaphorical distinction between "oil" variables, which never mix quadratically, and "vinegar" variables that facilitate the signing process.

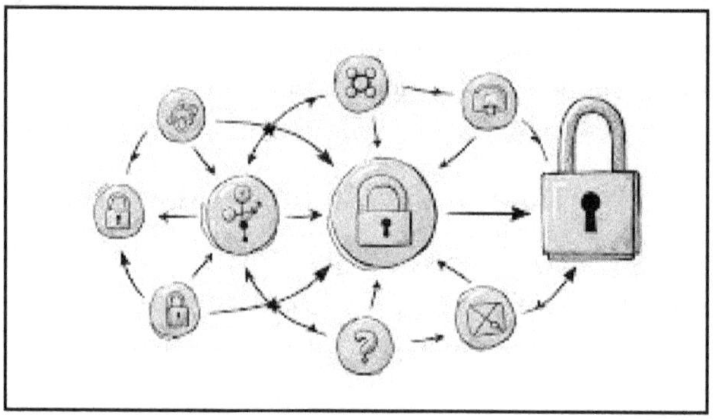

Fig: Vinegar (UOV): Exploring Multivariate Cryptography

The Unbalanced Oil and Vinegar signature scheme represents a robust and mathematically elegant approach to post-quantum digital signatures, grounded in the well-studied hardness of multivariate quadratic systems. While not selected as a primary standard in recent competitions, UOV continues to offer valuable properties—particularly its reliance on a single, well-understood computational problem and its relative implementation simplicity compared to some alternatives. The scheme's flexibility has spawned numerous variants addressing different security and efficiency requirements, demonstrating the vitality of multivariate cryptography research. As the cryptographic community navigates the complex transition to post-quantum security, UOV remains an important tool in the cryptographic toolbox, particularly for applications valuing conservative security assumptions or requiring alternatives to lattice-based constructions. Future advances in cryptanalysis, implementation techniques, and mathematical optimization will further clarify UOV's role in protecting communications against both classical and quantum adversaries.

Ongoing research continues to expand the capabilities and applications of UOV signatures. Investigations into shorter signature variants through carefully structured keys or modified polynomial systems aim to address one of the scheme's primary drawbacks. The development of identity-based and attribute-based UOV constructions could enable more advanced cryptographic protocols building on multivariate foundations. Another active area explores the combination of UOV with other post-quantum primitives in hybrid systems, providing defense-in-depth during the transition period. As quantum computing technology advances, the provable security of multivariate schemes against quantum adversaries will likely receive increased attention, potentially positioning UOV and its variants as long-term fallback options should other approaches encounter unexpected cryptanalysis.

Mathematical Foundations

The UOV scheme operates within the framework of multivariate quadratic (MQ) polynomials, leveraging the proven NP-hardness of solving random systems of quadratic equations. A typical UOV construction works over a finite field GF(q) with n variables divided into two distinct subsets: v vinegar variables and o oil variables (where $n = v + o$). The central map F consists of o quadratic polynomials in these n variables, constructed such that oil variables only appear linearly when vinegar variables are fixed.

This structural property enables efficient inversion of the map during signature generation while maintaining resistance against direct algebraic attacks. The public key P is formed by composing the central map F with two invertible affine transformations T and S ($P = T \circ F \circ S$), obscuring the underlying polynomial structure from potential attackers. The security of UOV relies fundamentally on the difficulty of decomposing P into its constituent parts without knowledge of the private key components (F, T, S).

Signature Generation and Verification Process

The UOV signature process demonstrates an elegant asymmetry between the computationally efficient signing operation and the verification procedure. To sign a message digest $h \in GF(q)^o$, the signer first selects random values for the v vinegar variables and substitutes them into the central map F. This reduces the o quadratic equations too linear equations in the remaining o oil variables, which can be solved efficiently using Gaussian elimination.

If the linear system proves inconsistent (which occurs with probability approximately $1/q$), new random vinegar values are selected until a solution is found. The complete solution (vinegar + oil values) then undergoes the affine transformation S to produce the final signature σ. Verification simply requires evaluating the public polynomials P at σ and checking for equality with h. This process ensures that while signature generation requires knowledge of the private key, verification can be performed by anyone with access to the public key, maintaining the essential properties of digital signature schemes.

Security Analysis and Known Attacks

The security of UOV signatures has been extensively analyzed against various cryptanalytic techniques, shaping the parameter choices for real-world implementations. The most direct attack approach involves solving the system of public quadratic equations using advanced algebraic techniques like Gröbner basis algorithms (e.g., F4/F5).

The complexity of these attacks grows exponentially with the vinegar-oil gap (v - o), leading to the security recommendation that v ≥ 2o. Other notable attack vectors include the reconciliation attack (effective against balanced Oil and Vinegar) and the minrank attack, both of which UOV's unbalanced structure resists. Current security estimates suggest that UOV with q=256, o=64, and v=128 provides approximately 128-bit post-quantum security, requiring signatures of about 10KB. The scheme's security remains intact against quantum computers because the underlying MQ problem resists known quantum algorithms like Shor's, though Grover's algorithm necessitates a modest increase in parameter sizes compared to classical security estimates.

Parameter Selection and Performance Considerations

Practical deployment of UOV requires careful balancing between security levels, signature sizes, and computational efficiency. The choice of finite field GF(q) significantly impacts both security and performance: while GF(2) offers compact representations, GF(256) provides better resistance against algebraic attacks and more efficient implementations on standard processors.

The vinegar-oil ratio v/o directly controls security against direct algebraic attacks, with v=2o being the conventional choice. However, recent analyses suggest slightly higher ratios (v ≈ 2.5o) may offer better long-term security margins. Computational performance is dominated by the matrix operations during signing and verification, with optimized implementations leveraging SIMD instructions for finite field arithmetic. On modern CPUs, UOV can generate thousands of signatures per second, though verification remains more computationally intensive due to the need to evaluate all public polynomials. These trade-offs make UOV particularly suitable for applications where signing occurs on constrained devices while verification happens on more powerful infrastructure.

Variants and Enhancements to Basic UOV

Several modifications to the basic UOV scheme have been proposed to address specific security or efficiency concerns. The Rainbow signature scheme, a multilayer variant of UOV, reduces signature sizes by introducing intermediate "vinegar" variables between layers. Cyclic UOV incorporates structured public keys to minimize storage requirements, though this approach requires careful security analysis against potential exploitable symmetries. Projective UOV modifies the central map to resist the latest rank-based attacks while maintaining comparable efficiency. Another important direction is the development of threshold UOV schemes that enable distributed signature generation, enhancing security in scenarios where private key protection is paramount. These variants demonstrate the flexibility of the UOV framework while highlighting the ongoing need for cryptanalysis as new mathematical techniques emerge.

Standardization Status and Practical Adoption

UOV and its variants have been actively considered in the NIST Post-Quantum Cryptography Standardization process, though they were not selected for final standardization in Round 3. Despite this, UOV remains an important candidate for several reasons. Its conservative security foundations (based solely on the MQ problem without additional assumptions) make it attractive for long-term security applications. The scheme's relatively simple implementation requirements compared to lattice-based alternatives have led to adoption in certain niche applications, particularly in embedded systems where code size constraints favor multivariate schemes. Ongoing research into improved UOV variants continues within the cryptographic community, with potential for future inclusion in supplementary standards or industry-specific protocols. The Internet Engineering Task Force (IETF) has included UOV in draft standards for post-quantum signatures, ensuring interoperability guidelines exist for potential implementers.

Comparison with Other Post-Quantum Signature Schemes

When evaluated against other post-quantum signature candidates, UOV presents a distinctive set of characteristics. Compared to lattice-based schemes like Dilithium, UOV offers simpler arithmetic operations but larger signature sizes (typically 5-10x larger for comparable security levels). Against hash-based schemes such as SPHINCS+, UOV provides significantly faster signing and verification but requires state management to prevent key reuse. The multivariate approach avoids the reliance on number-theoretic problems that make lattice and code-based schemes vulnerable to future mathematical advances, though this comes at the cost of less compact representations. For applications prioritizing implementation simplicity and resistance to quantum attacks without additional mathematical assumptions, UOV remains a compelling choice despite not being selected as a primary NIST standard.

Implementation Challenges and Side-Channel Resistance

Practical implementation of UOV introduces several engineering challenges that must be addressed for secure deployment. The large key sizes (often hundreds of kilobytes for public keys) require careful memory management in constrained environments. Finite field arithmetic implementations must maintain constant-time execution to prevent timing side-channel attacks, particularly during the Gaussian elimination steps in signature generation. Fault injection attacks could potentially recover private key information by manipulating intermediate values during signing operations, necessitating appropriate countermeasures in security-critical implementations.

The substantial difference between signing and verification complexity also creates potential denial-of-service vectors that protocol designers must consider. These factors have motivated research into hardware acceleration approaches, including FPGA implementations that can handle the scheme's large matrix operations efficiently while maintaining physical security.

4.4 HASH-BASED SIGNATURES

4.4.1 SPHINCS+

SPHINCS$^+$, also known officially as SLH-DSA, is a post-quantum signature scheme selected by the NIST for the FIPS 205 standard of the post-quantum standardisation process.

Design

SPHINCS$^+$ is based on a one-time signature scheme called WOTS$^+$ (a modified version of the Winternitz one-time signature scheme), a few-time signature scheme called FORS (Forest of Random Subsets) and merkle trees.

When signing, the message is signed with a FORS key. The FORS key is signed with a WOTS$^+$ key that is a leaf of a merkle tree. The root of the tree is then signed with another WOTS$^+$ key that is itself a leaf of another tree. That tree's root is again signed with a WOTS$^+$. The number of layers of trees is a parameter that is specified as part of the algorithm. This "tree of trees" is called a hypertree. The root of the top tree is the public key. The signature consists of the FORS key and its signature, the WOTS$^+$ keys with their signatures and inclusion proofs for the merkle tree and a random value called R that was used to generate the path in the hypertree.

In order to verify a signature, the verifier first verifies the first WOTS$^+$ key's inclusion proof against the public key and then verifies the key's signature of the next root. Then, they check the next WOTS$^+$ key's inclusion proof against the new root. This goes on until the last WOTS$^+$ key is reached, which is then used to verify the FORS key. That key is then used to actually verify the message's signature.

All WOTS$^+$ keys and FORS keys are generated deterministically from the private key. During signing, the signer generates a random bit string called R and hashes it together with the message. Parts of the resulting hash are used to select the path through the hypertree while the rest is signed with the FORS key.

Security

SPHINCS$^+$ has been called a "conservative" choice by NIST since its security solely relies on the preimage and collision resistance of the underlying hash function.

Instances

SLH-DSA specifies the following instances based on the hash function SHA256 or SHAKE256), the type (f for faster signing time and s for shorter signature) and security level (e.g. 128 means that forging signatures is as hard as breaking AES-128):

Name	Security level	Type	Hash function	Public key size	Private key size	Signature size
SPHINCS+-SHA2-128s	1	small	SHA256	32	64	7856
SPHINCS+-SHAKE-128s			SHAKE256			
SPHINCS+-SHA2-128f		fast	SHA256			17088
SPHINCS+-SHAKE-128f			SHAKE256			
SPHINCS+-SHA2-192s	3	small	SHA256	48	96	16224
SPHINCS+-SHAKE-192s			SHAKE256			
SPHINCS+-SHA2-192f		fast	SHA256			35664
SPHINCS+-SHAKE-192f			SHAKE256			
SPHINCS+-SHA2-256s	5	small	SHA256	64	128	29792
SPHINCS+-SHAKE-256s			SHAKE256			
SPHINCS+-SHA2-256f		fast	SHA256			49856
SPHINCS+-SHAKE-256f			SHAKE256			

4.4.2 XMSS

In the rapidly evolving field of cryptography, hash-based schemes have emerged as one of the most reliable alternatives to traditional public key systems in the face of quantum threats. Unlike number-theoretic cryptography, which relies on the hardness of problems like integer factorization or elliptic curve discrete logarithms, hash-based cryptography is built on the foundation of cryptographic hash functions. These functions are considered quantum-resistant since no efficient quantum algorithms are known to significantly weaken their security. The Extended Merkle Signature Scheme, commonly referred to as XMSS, is a notable development in this domain. XMSS is designed specifically to provide forward secure and post-quantum secure digital signatures while being stateful. It was the first hash-based signature scheme to be standardized by the IETF and NIST, highlighting its practical relevance in modern cryptographic infrastructure.

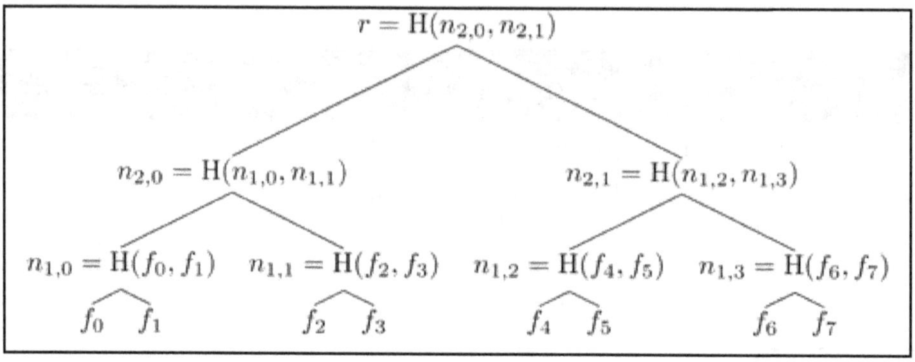

Fig: XMSS

XMSS represents a significant milestone in the evolution of cryptographic signatures, offering a practical, quantum-resistant solution based solely on the strength of cryptographic hash functions. Its reliance on well-understood primitives, its resilience to quantum attacks, and its formal standardization make it an attractive choice for many security-conscious applications. While its statefulness presents implementation challenges, these can be managed effectively with proper engineering practices. As digital security becomes increasingly important and quantum computing looms as a realistic threat, schemes like XMSS will play a crucial role in safeguarding the integrity and authenticity of digital communications. Its adoption signals a shift towards more resilient cryptographic systems, ensuring that the foundations of trust in the digital world remain secure in the quantum age.

Research on XMSS continues to evolve, with efforts directed at improving efficiency, reducing signature sizes, and making state management more reliable. One promising direction is the development of hybrid schemes that combine hash-based and other post-quantum techniques to achieve a better balance between performance and usability. Additionally, work is being done to optimize XMSS for constrained environments such as embedded systems and smart cards. Techniques such as Winternitz one-time signatures, which are used within XMSS to reduce key sizes and increase signing speed, are also being refined to enhance overall efficiency. Other areas of active research include secure and fault-tolerant implementations, especially in systems where power loss or hardware errors could lead to accidental key reuse. With the threat of quantum computing on the horizon, continued focus on robust and efficient hash-based signatures like XMSS is likely to be a central pillar of cryptographic research and deployment in the coming years.

Conceptual Foundation of XMSS

XMSS is grounded in the principles of Merkle trees and one-time signature schemes. Merkle trees are data structures that allow many one-time public keys to be compressed into a single root value, which serves as the public key of the entire signature system. XMSS uses this tree structure to aggregate a large number of one-time keys, which are then used one after another to produce digital signatures. Each signature reveals part of the structure, hence the system is inherently stateful. This stateful property means that the system must carefully manage which keys have been used to prevent reuse, which would compromise security. XMSS achieves a balance between security and efficiency, and its use of hash functions as the only cryptographic primitive makes it a strong candidate for systems seeking resistance against both classical and quantum adversaries.

Structure and Operation of XMSS

The operation of XMSS begins with the generation of a set of one-time key pairs, each to be used for a single digital signature. These keys are then arranged in a binary Merkle tree, with the leaves representing the public keys of the one-time signature scheme. To sign a message, the scheme selects the next unused one-time key, uses it to sign the message, and includes the authentication path from the corresponding leaf to the root. This allows a verifier to confirm the integrity of the one-time public key and thus validate the signature. Each signature includes the message, the one-time signature, and the corresponding authentication path. Since each one-time key must only be used once, the signer must maintain a state indicating which keys have been used. While this statefulness introduces complexity in implementation, it also guarantees strong forward security, meaning that compromise of the current state does not endanger previously generated signatures.

Security Assumptions and Quantum Resistance

The security of XMSS is based on the collision resistance and pre-image resistance properties of the underlying hash function. As of current knowledge, Grover's algorithm, a quantum algorithm for unstructured search, can only provide a quadratic speed-up against hash functions. This means that for a hash function with n-bit output, a quantum adversary would need 2^(n/2) operations to break it, which is still computationally expensive with large n. Consequently, XMSS offers robust post-quantum security without relying on hard algebraic problems that may be rendered obsolete by Shor's algorithm or other quantum innovations. The simplicity and well-understood nature of hash functions also reduce the attack surface, making XMSS a very conservative and trustworthy choice in cryptographic applications. Unlike lattice-based or multivariate schemes that are newer and potentially more complex, XMSS builds on decades of research and practice with hash functions, lending confidence to its security claims.

Stateless vs. Stateful Designs and XMSS's Position

One of the primary trade-offs in XMSS is its statefulness. Each time a message is signed, the scheme must update its internal state to record that a particular one-time key has been used. Reuse of keys leads to catastrophic failure, allowing adversaries to forge signatures. This makes implementation of XMSS more challenging, particularly in distributed systems or environments where power loss or system resets can occur. To mitigate this, various strategies such as state backup, redundant state storage, and secure counters are employed. In contrast, stateless hash-based signature schemes like SPHINCS+ remove this requirement by sacrificing some performance and increasing the size of signatures. Nevertheless, XMSS offers a favorable trade-off for many applications, providing compact signatures and faster signing speeds in exchange for careful state management. The statefulness, while a burden, enables a much smaller signature size and more efficient operations compared to stateless alternatives.

Applications and Practical Deployment of XMSS

XMSS has seen increasing interest for deployment in systems where long-term security and quantum resistance are essential. One such domain is firmware signing, where ensuring the authenticity and integrity of updates is critical for secure device operation. Given the limited number of updates typically required over the lifetime of a device, the one-time key limitation of XMSS is not a significant concern. Furthermore, its forward-secure nature is ideal for scenarios where old signatures must remain verifiable even if the current signing key is compromised. XMSS is also being explored for use in secure boot processes, IoT devices, and national security systems, where robustness against future attacks is essential. Its inclusion in cryptographic libraries and its acceptance by international standards bodies have paved the way for broader adoption, particularly in applications where the complexity of managing state can be justified by the enhanced security benefits.

Standardization and Support by Cryptographic Bodies

The formal recognition of XMSS by major standards organizations has further solidified its standing in the cryptographic community. It is defined in RFC 8391 by the Internet Engineering Task Force (IETF), providing detailed specifications for implementation and integration. XMSS is also under consideration by the National Institute of Standards and Technology (NIST) in its post-quantum cryptography standardization process. Although NIST has prioritized stateless schemes like SPHINCS+ for broader usability, the importance of having multiple paradigms with different security assumptions is well-recognized. The inclusion of XMSS ensures diversity in cryptographic infrastructure and offers a strong fallback option should unforeseen weaknesses arise in other schemes. Additionally, the structured yet conservative design of XMSS makes it an excellent pedagogical tool for understanding hash-based cryptography and signature systems in general.

4.4.3 LMS

In the rapidly changing landscape of cryptographic security, hash-based cryptography has emerged as one of the most promising approaches for post-quantum security. Among these schemes, the Leighton-Micali Signature (LMS) system stands out as a particularly robust and well-structured solution for digital signatures. As quantum computing threatens to break classical public-key cryptography, hash-based approaches like LMS offer a compelling alternative that relies on the security of cryptographic hash functions rather than number-theoretic assumptions vulnerable to quantum attacks.

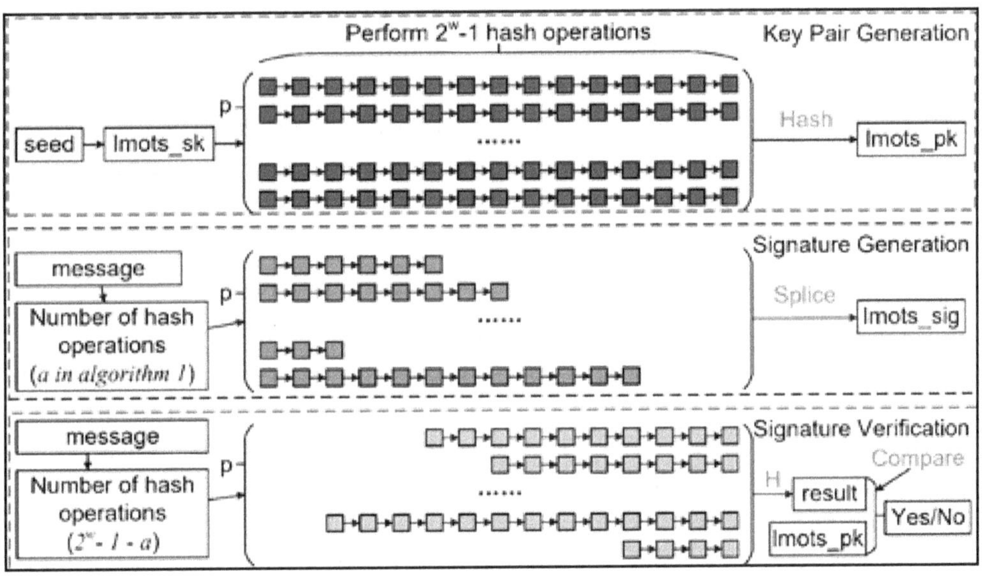

Fig: Leighton-Micali Signature

The Leighton-Micali Signature scheme represents a significant advancement in hash-based cryptography, offering a robust and efficient approach to post-quantum digital signatures. Its hierarchical structure and reliance on well-studied cryptographic hash functions provide strong security guarantees while maintaining practical performance characteristics. As one of the most well-understood post-quantum signature schemes, LMS demonstrates the potential of hash-based approaches to secure digital communications in an era where traditional cryptographic methods may no longer be sufficient. While challenges remain in terms of signature sizes and state management, ongoing research continues to address these issues. The mathematical foundations of hash-based cryptography, combined with the practical improvements embodied in LMS, position this approach as an important component of the post-quantum cryptographic landscape. As the field evolves, schemes like LMS will play a crucial role in developing the quantum-resistant cryptographic systems of tomorrow.

The field of hash-based cryptography continues to evolve, with LMS serving as an important milestone in this development. Future research directions include exploring new tree structures and key management techniques that could offer improved security or efficiency characteristics. Investigations into alternative hash function constructions or usage paradigms may lead to schemes with better performance profiles while maintaining strong security guarantees. Additionally, the study of new cryptanalytic techniques will be crucial for understanding the long-term security of these schemes and guiding parameter selection.

Another promising area of research involves hybrid approaches that combine hash-based cryptography with other post-quantum primitives. These combinations could potentially leverage the strengths of multiple cryptographic approaches to achieve optimal security and efficiency trade-offs. The development of more efficient algorithms for hash function operations could also significantly improve the practicality of hash-based schemes. As quantum computing advances, the continued evolution of hash-based cryptography will be essential for maintaining secure digital communications in the post-quantum era.

Hash-based cryptography represents a fundamentally different approach to digital signatures compared to traditional number-theoretic methods. Instead of relying on the hardness of mathematical problems like integer factorization or discrete logarithms, hash-based schemes derive their security from the properties of cryptographic hash functions. These functions, such as SHA-256 or SHA-3, are designed to be one-way, collision-resistant, and preimage-resistant, making them suitable building blocks for secure cryptographic constructions. The security of hash-based signatures depends on the assumption that finding collisions or preimages for these hash functions remains computationally infeasible, even for quantum computers.

The basic principle behind hash-based signatures involves using hash functions to create one-time signature keys that are used only once before being discarded. This approach, known as the one-time signature (OTS) paradigm, provides information-theoretic security for each individual signature while maintaining computational efficiency. By combining multiple OTS instances with a tree-based structure, hash-based signature schemes can achieve practical security with reasonable key sizes and computational requirements. This foundation forms the basis for more sophisticated constructions like LMS, which extend these principles to create efficient and secure digital signature systems.

Leighton-Micali Signature Scheme

The Leighton-Micali Signature (LMS) scheme represents a significant advancement in hash-based cryptography by introducing a hierarchical structure that improves efficiency while maintaining strong security guarantees. Developed by Thomas Leighton and Silvio Micali, this system builds upon earlier hash-based signature concepts to create a more practical and scalable solution. LMS organizes one-time signature keys in a tree structure, where each node represents a hash of its child nodes, allowing for efficient verification and key management. This hierarchical approach enables the scheme to support a large number of signatures while keeping the public key size manageable.

The LMS scheme operates by first generating a tree of hash values, where the root of the tree serves as the public key. To sign a message, the signer uses a one-time signature key from a leaf node, along with authentication paths that allow the verifier to reconstruct the public key. This process ensures that each signature is unique and that the security of the system depends on the collision resistance of the underlying hash function. The tree structure allows for efficient signature generation and verification, making LMS more practical for real-world applications compared to earlier hash-based signature schemes.

Security Properties and Cryptanalysis

The security of the LMS scheme is based on several well-established cryptographic principles that contribute to its resistance against known attacks. The primary security assumption is the collision resistance of the underlying hash function, which ensures that an attacker cannot find two different inputs that produce the same hash output. This property is crucial for preventing forgery attacks, as creating a valid signature would require finding such a collision. Additionally, the one-time nature of the signature keys means that even if a particular key is compromised, it cannot be reused to forge other signatures, providing a strong security guarantee.

Cryptanalytic research on LMS has generally confirmed its resistance to known attacks when properly implemented. The security of the scheme depends on the careful selection of parameters, including the height of the tree structure and the size of the hash function outputs. These parameters must be chosen to balance security and efficiency, ensuring that the scheme remains secure against brute-force and collision-finding attacks. The hierarchical structure of LMS provides additional security by limiting the impact of any single compromised key, as the tree structure contains the damage to specific branches.

Performance Characteristics and Practical Considerations

One of the most significant advantages of the LMS scheme is its computational efficiency, particularly in verification operations. Unlike many post-quantum signature schemes that involve complex mathematical operations, LMS relies primarily on hash function evaluations, which are generally fast and efficient. This makes the scheme particularly well-suited for applications where verification speed is critical, such as in high-throughput systems or resource-constrained environments. The signature generation process is also efficient, though it requires careful management of the one-time signature keys to maintain security.

However, the LMS scheme does face some practical limitations that must be considered in real-world deployments. The most notable of these is the relatively large signature sizes compared to classical digital signature schemes like RSA or ECDSA. Each signature in LMS includes not only the one-time signature itself but also the authentication path through the tree structure, which can result in larger overall signature sizes. Additionally, the stateful nature of the scheme, where each one-time key can only be used once, requires careful key management to prevent reuse and ensure security. These considerations highlight the trade-offs involved in hash-based cryptography and the importance of proper implementation.

Parameter Selection and Security Trade-offs

The security and efficiency of the LMS scheme are heavily dependent on the careful selection of its parameters, which determine the structure and security of the signature system. These parameters include the height of the tree structure, the size of the hash function outputs, and the number of one-time signature keys available. Larger parameters generally provide stronger security by increasing the computational effort required for collision-finding attacks but result in larger signature sizes and greater computational requirements. The choice of these parameters involves complex trade-offs between security, signature size, and computational efficiency.

The process of parameter selection for LMS involves both theoretical analysis and practical experimentation to ensure resistance against known attacks while maintaining acceptable performance. Cryptographers must consider the latest developments in cryptanalysis when choosing parameters, as new attack techniques may require adjustments to maintain security. This dynamic relationship between parameter selection and security analysis highlights the importance of ongoing research in hash-based cryptography. The flexibility of LMS in parameter choices allows it to be adapted to different security requirements, making it a versatile option for various applications.

Comparison with Other Post-Quantum Signature Schemes

When evaluating LMS in the context of other post-quantum digital signature schemes, several comparative advantages and disadvantages become apparent. Compared to lattice-based signature schemes, LMS offers potentially stronger security guarantees based on well-studied hash function properties rather than relatively newer mathematical constructs.

The security of hash-based schemes is often more straightforward to analyze, as it relies on the collision resistance of hash functions that have been extensively studied and standardized. However, lattice-based schemes often provide better performance characteristics, particularly in terms of signature sizes and computational efficiency.

Compared to multivariate signature schemes, LMS offers a different balance of security and efficiency. Multivariate schemes typically provide smaller signature sizes but may have more complex security assumptions and potentially larger key sizes. The choice between these approaches often depends on the specific requirements of the application, with hash-based schemes like LMS providing a compelling option when security based on well-understood primitives is prioritized. The diversity of post-quantum approaches ensures that different security needs can be addressed, with hash-based schemes playing an important role in this cryptographic ecosystem.

Implementation Challenges and Considerations

Implementing LMS in practical systems requires careful attention to several factors to ensure both security and performance. The stateful nature of the scheme, where each one-time signature key must be used only once, demands robust key management to prevent reuse and maintain security. This requires careful tracking of key usage and proper implementation of the tree structure to ensure that signatures remain valid and secure. Additionally, the relatively large signature sizes characteristic of hash-based schemes must be managed efficiently, particularly in environments where bandwidth or storage is constrained.

Another implementation consideration is the need for secure and efficient hash function evaluations. While modern cryptographic hash functions are generally fast and efficient, their implementation must be carefully optimized to ensure performance while maintaining security. Side-channel resistance is also an important consideration, as timing attacks or other physical observations could potentially leak information about the signing process. Constant-time implementations and other countermeasures are essential to protect against these potential vulnerabilities, ensuring that the theoretical security of LMS translates into practical protection.

Standardization Efforts and Cryptographic Evaluation

The LMS scheme has been recognized as a strong candidate in the ongoing efforts to standardize post-quantum cryptographic algorithms. The National Institute of Standards and Technology (NIST) included LMS in its Post-Quantum Cryptography Standardization Project, subjecting it to rigorous evaluation alongside other promising schemes. This evaluation process involves extensive cryptanalysis, performance testing, and implementation scrutiny to assess the scheme's suitability for standardization. The inclusion of LMS in these efforts highlights its potential contributions to post-quantum security, particularly in applications where hash-based signatures offer advantages.

The cryptographic community continues to study and refine hash-based signature schemes like LMS, recognizing their potential to provide secure digital signatures in the post-quantum era. Ongoing research focuses on addressing the practical limitations of these schemes while maintaining their strong security properties. The lessons learned from the standardization process will inform future developments in hash-based cryptography, potentially leading to new variants or improved implementations that could play a role in future cryptographic standards.

4.5 ISOGENY-BASED CRYPTOGRAPHY

4.5.1 SIDH

Supersingular isogeny Diffie–Hellman key exchange (SIDH or SIKE) is an insecure proposal for a post-quantum cryptographic algorithm to establish a secret key between two parties over an untrusted communications channel. It is analogous to the Diffie–Hellman key exchange, but is based on walks in a supersingular isogeny graph and was designed to resist cryptanalytic attack by an adversary in possession of a quantum computer. Before it was broken, SIDH boasted one of the smallest key sizes of all post-quantum key exchanges; with compression, SIDH used 2688-bit public keys at a 128-bit quantum security level.

SIDH also distinguishes itself from similar systems such as NTRU and Ring-LWE by supporting perfect forward secrecy, a property that prevents compromised long-term keys from compromising the confidentiality of old communication sessions. These properties seemed to make SIDH a natural candidate to replace Diffie–Hellman (DHE) and elliptic curve Diffie–Hellman (ECDHE), which are widely used in Internet communication. However, SIDH is vulnerable to a devastating key-recovery attack published in July 2022 and is therefore insecure. The attack does not require a quantum computer. For certain classes of problems, algorithms running on quantum computers are naturally capable of achieving lower time complexity than on classical computers. That is, quantum algorithms can solve certain problems faster than the most efficient algorithm running on a traditional computer. For example, Shor's algorithm can factor an integer N in polynomial time, while the best-known factoring classic algorithm, the general number field sieve, operates in sub-exponential time.

This is significant to public key cryptography because the security of RSA is dependent on the infeasibility of factoring integers, the integer factorization problem. Shor's algorithm can also efficiently solve the discrete logarithm problem, which is the basis for the security of Diffie–Hellman, elliptic curve Diffie–Hellman, elliptic curve DSA, Curve25519, ed25519, and ElGamal. Although quantum computers are currently in their infancy, the ongoing development of quantum computers and their theoretical ability to compromise modern cryptographic protocols (such as TLS/SSL) has prompted the development of post-quantum cryptography.

SIDH was created in 2011 by De Feo, Jao, and Plut. It uses conventional elliptic curve operations and is not patented. SIDH provides perfect forward secrecy and thus does not rely on the security of long-term private keys. Forward secrecy improves the long-term security of encrypted communications, helps defend against mass surveillance, and reduces the impact of vulnerabilities like Heartbleed.

Security

The most straightforward way to attack SIDH is to solve the problem of finding an isogeny between two supersingular elliptic curves with the same number of points. At the time of the original publication due to De Feo, Jao and Plût, the best attack known against SIDH was based on solving the related claw-finding problem, which led to a complexity of $O(p^{1/4})$ for classical computers and $O(p^{1/6})$ for quantum computers.

This suggested that SIDH with a 768-bit prime (p) would have a 128-bit security level. A 2014 study of the isogeny problem by Delfs and Galbraith confirmed the $O(p^{1/4})$ security analysis for classical computers. The classical security $O(p^{1/4})$ remained unaffected by related cryptanalytic work of Biasse, Jao and Sankar as well as Galbraith, Petit, Shani and Yan.

A more intricate attack strategy is based on exploiting the auxiliary elliptic-curve points present in SIDH public keys, which in principle reveal a lot of additional information about the secret isogenies, but this information did not seem computationally useful for attackers at first. Petit in 2017 first demonstrated a technique making use of these points to attack some rather peculiar SIDH variants. Despite follow-up work extending the attack to much more realistic SIDH instantiations, the attack strategy still failed to break "standard" SIDH as employed by the NIST PQC submission SIKE.

In July 2022, Castryck and Decru published an efficient key-recovery attack on SIKE that exploits the auxiliary points. Using a single-core computer, SIKEp434 was broken within approximately an hour, SIKEp503 within approximately 2 hours, SIKEp610 within approximately 8 hours and SIKEp751 within approximately 21 hours. The attack relies on gluing together multiple of the elliptic curves appearing in the SIDH construction, giving an abelian surface (more generally, an abelian variety), and computing a specially crafted isogeny defined by the given auxiliary points on that higher-dimensional object.

It should be stressed that the attack crucially relies on the auxiliary points given in SIDH, and there is no known way to apply similar techniques to the general isogeny problem.

Efficiency

During a key exchange, entities A and B will each transmit information of 2 coefficients modulo p^2 defining an elliptic curve and 2 elliptic curve points. Each elliptic curve coefficient requires $\log_2 p^2$ bits. Each elliptic curve point can be transmitted in $1 + \log_2 p^2$ bits; hence, the transmission is $4 + 4 \log_2 p^2$ bits. This is 6144 bits for a 768-bit modulus p (128-bit security).

However, this can be reduced by over half to 2640 bits (330 bytes) using key-compression techniques, the latest of which appears in recent work by authors Costello, Jao, Longa, Naehrig, Renes and Urbanik. With these compression techniques, SIDH has a similar bandwidth requirement to traditional 3072-bit RSA signatures or Diffie-Hellman key exchanges. This small space requirement makes SIDH applicable to context that have a strict space requirement, such as Bitcoin or Tor. Tor's data cells must be less than 517 bytes in length, so they can hold 330-byte SIDH keys. By contrast, NTRUEncrypt must exchange approximately 600 bytes to achieve a 128-bit security and cannot be used within Tor without increasing the cell size.

In 2014, researchers at the University of Waterloo developed a software implementation of SIDH. They ran their partially optimized code on an x86-64 processor running at 2.4 GHz. For a 768-bit modulus they were able to complete the key exchange computations in 200 milliseconds thus demonstrating that the SIDH is computationally practical.

In 2016, researchers from Microsoft posted software for the SIDH which runs in constant time (thus protecting against timing attacks) and is the most efficient implementation to date. They write: "The size of public keys is only 564 bytes, which is significantly smaller than most of the popular post-quantum key exchange alternatives. Ultimately, the size and speed of our software illustrates the strong potential of SIDH as a post-quantum key exchange candidate and we hope that these results encourage a wider cryptanalytic effort." The code is open source (MIT) and is available on GitHub: https://github.com/microsoft/PQCrypto-SIDH.

In 2016, researchers from Florida Atlantic University developed efficient ARM implementations of SIDH and provided a comparison of affine and projective coordinates. In 2017, researchers from Florida Atlantic University developed the first FPGA implementations of SIDH.

Similar Systems, Signatures and Uses

A predecessor to the SIDH was published in 2006 by Rostovtsev and Stolbunov. They created the first Diffie-Hellman replacement based on elliptic curve isogenies. Unlike the method of De Feo, Jao, and Plut, the method of Rostovtsev and Stolbunov used ordinary elliptic curves and was found to have a subexponential quantum attack.

In March 2014, researchers at the Chinese State Key Lab for Integrated Service Networks and Xidian University extended the security of the SIDH to a form of digital signature with strong designated verifier. In October 2014, Jao and Soukharev from the University of Waterloo presented an alternative method of creating undeniable signatures with designated verifier using elliptic curve isogenies.

4.5.2 SIKE, and the Recent Break

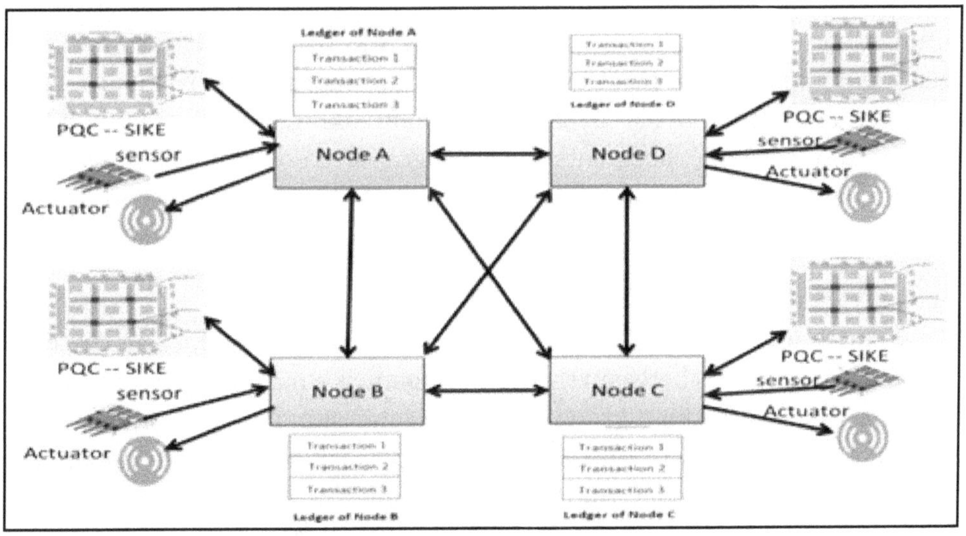

Fig: Optimizing SIKE for Blockchain-Based IoT Ecosystems with Resource Constraints

In the competitive landscape of post-quantum cryptography, Supersingular Isogeny Key Encapsulation (SIKE) emerged as one of the most mathematically sophisticated and promising candidates for quantum-resistant encryption. Developed through years of academic research, SIKE represented a fundamentally different approach to cryptographic security by leveraging the complex mathematics of elliptic curve isogenies. Unlike other post-quantum candidates that rely on lattice problems, hash functions, or multivariate equations, SIKE's security was based on the hardness of computing isogenies between supersingular elliptic curves—a problem believed to be resistant to both classical and quantum attacks. This innovative approach positioned SIKE as a strong contender in the National Institute of Standards and Technology's (NIST) post-quantum cryptography standardization process, ultimately reaching the final round of evaluation.

The story of SIKE represents a fascinating chapter in the development of post-quantum cryptography, illustrating both the promise and challenges inherent in this rapidly evolving field. As one of the most mathematically sophisticated post-quantum candidates, SIKE captured the imagination of the cryptographic community with its elegant construction and strong theoretical foundations. The eventual discovery of its vulnerabilities, while disappointing for its proponents, ultimately served to strengthen the cryptographic standardization process by demonstrating its effectiveness in identifying weaknesses.

The legacy of SIKE extends beyond its specific protocol to the broader understanding of cryptographic security and the standardization process. The experience highlighted the importance of continuous cryptanalysis, conservative parameter selection, and the need for cryptographic agility in an evolving security landscape. As the field moves forward with the standardization of other post-quantum algorithms, the lessons learned from SIKE will continue to inform the development and evaluation of new cryptographic approaches. The mathematical insights gained through the study and eventual breaking of SIKE will undoubtedly contribute to future advancements in cryptographic research, ensuring that the legacy of this innovative protocol endures even as its specific construction fades from consideration.

Mathematical Foundations of SIKE

SIKE's security was rooted in the intricate mathematics of supersingular elliptic curves and the isogenies between them. An isogeny is a morphism between elliptic curves that preserves their group structure, and supersingular elliptic curves represent a special class of elliptic curves with particularly rich mathematical properties. The core computational problem underlying SIKE's security was the Supersingular Isogeny Diffie-Hellman (SIDH) problem, which involves computing an isogeny between two supersingular curves given only their public representations. This problem was believed to be exceptionally hard, even for quantum computers, due to the lack of known efficient algorithms for solving it. The SIKE protocol operated by having two parties exchange public keys derived from isogenies of supersingular elliptic curves. These keys were constructed in such a way that computing the shared secret required solving the SIDH problem, which was conjectured to be computationally infeasible. The elegance of this approach lay in its ability to provide key exchange functionality similar to classical elliptic curve cryptography but with security based on fundamentally different mathematical assumptions. This mathematical novelty made SIKE particularly attractive as a post-quantum candidate, as it offered a fresh perspective on cryptographic hardness that appeared resistant to known quantum attack strategies.

SIKE's Journey through Standardization

SIKE's path through the NIST post-quantum cryptography standardization process was marked by both admiration for its mathematical sophistication and concerns about its practical aspects. As one of the finalists in NIST's evaluation, SIKE underwent rigorous scrutiny from cryptanalysts worldwide, with researchers examining its security assumptions, implementation characteristics, and potential vulnerabilities. The protocol's inclusion in the final round of evaluation reflected the cryptographic community's recognition of its innovative approach and the theoretical strength of its underlying mathematical problems.

Throughout the standardization process, SIKE demonstrated several attractive features that distinguished it from other post-quantum candidates. Its relatively small key sizes compared to some lattice-based alternatives made it appealing for applications where bandwidth was a concern. Additionally, the mathematical structure of isogeny-based cryptography offered potential for efficient implementations, particularly in hardware. These characteristics positioned SIKE as a strong candidate for standardization, with many in the cryptographic community hopeful that it might become part of the next generation of cryptographic standards.

Breakthrough Attack on SIKE

The cryptographic community's confidence in SIKE was dramatically challenged by a groundbreaking attack published in 2022. Researchers Wouter Castryck and Thomas Decru developed an efficient classical algorithm that could solve the SIDH problem at the heart of SIKE's security. This attack exploited a previously unknown mathematical relationship between supersingular isogeny graphs and the theory of torsion point graphs, allowing for the computation of isogenies in polynomial time.

The discovery represented a significant theoretical advancement in computational number theory while simultaneously undermining the security assumptions upon which SIKE was built. The attack demonstrated that the computational problems underlying SIKE were not as hard as previously believed, particularly when considering the specific parameters used in the protocol. While the attack did not immediately break all variants of SIKE, it showed that the security margins were significantly smaller than originally estimated. This revelation had immediate consequences for SIKE's viability as a cryptographic standard, as it called into question the long-term security guarantees that were essential for its adoption. The attack's publication marked a turning point in the evaluation of SIKE, leading to its eventual exclusion from NIST's final post-quantum cryptography standards.

Implications of the SIKE Break

The successful attack on SIKE had profound implications for both the specific protocol and the broader field of post-quantum cryptography. For SIKE itself, the attack demonstrated that despite its elegant mathematical foundations, the protocol's security assumptions were not as robust as initially believed. This realization underscored the importance of conservative parameter selection in cryptographic design, as the attack showed that even theoretically sound mathematical problems can have practical vulnerabilities when instantiated in specific ways.

More broadly, the SIKE break highlighted the dynamic nature of cryptographic research and the challenges inherent in post-quantum standardization efforts. The attack served as a reminder that cryptographic security is an evolving field where new mathematical insights can dramatically alter the security landscape. This event reinforced the need for ongoing cryptanalysis and the importance of maintaining healthy skepticism even about well-established cryptographic constructions. The cryptographic community's response to this development demonstrated the resilience of the standardization process, which is designed to identify and address such vulnerabilities before finalizing standards.

Lessons Learned from the SIKE Experience

The rise and fall of SIKE as a post-quantum cryptographic contender offers several valuable lessons for the cryptographic community. One of the most significant takeaways is the importance of thorough and continuous cryptanalysis throughout the standardization process. The SIKE break demonstrated that even mathematically elegant constructions with strong theoretical foundations can harbor unexpected vulnerabilities. This underscores the necessity of subjecting cryptographic proposals to extensive scrutiny from diverse research perspectives.

Another important lesson is the value of conservative parameter selection in cryptographic design. The attack on SIKE showed that the specific instantiation of a cryptographic primitive can significantly impact its security, even when the underlying mathematical problem is believed to be hard. This highlights the need for cryptographic parameters to be chosen with substantial security margins to account for potential future advances in cryptanalysis. The SIKE experience also illustrated the benefits of having multiple cryptographic approaches in development, as the diversity of post-quantum candidates helps ensure that the failure of one does not leave the field without alternatives.

Future of Isogeny-Based Cryptography

While the specific SIKE protocol may no longer be viable as a cryptographic standard, the attack does not necessarily spell the end for isogeny-based cryptography as a whole. The mathematical richness of supersingular elliptic curves and their isogenies continues to offer potential for cryptographic constructions. Researchers are already exploring alternative isogeny-based protocols that might avoid the vulnerabilities identified in the SIKE attack. These efforts include investigating different parameter sets, alternative mathematical constructions, and new cryptographic primitives that could leverage isogenies in more secure ways.

The cryptographic community's response to the SIKE break has been characterized by both disappointment at the loss of a promising candidate and excitement about the mathematical advancements that enabled the attack. This dual perspective reflects the dynamic nature of cryptographic research, where breakthroughs in understanding can simultaneously undermine existing constructions while opening new avenues for exploration. The future of isogeny-based cryptography will likely involve a more nuanced understanding of the computational hardness of these mathematical problems, potentially leading to new cryptographic schemes that build upon the lessons learned from SIKE.

Broader Implications for Post-Quantum Cryptography

The SIKE experience offers important insights for the broader field of post-quantum cryptography. The event demonstrated the critical role of the standardization process in identifying and addressing vulnerabilities before cryptographic algorithms are widely deployed. The rigorous evaluation that SIKE underwent as part of NIST's process ultimately served its purpose by revealing the protocol's weaknesses before it could be adopted as a standard. This outcome validates the importance of comprehensive cryptanalysis in the development of new cryptographic standards.

The attack on SIKE also highlighted the need for cryptographic agility—the ability to transition between cryptographic algorithms as new vulnerabilities are discovered. As the cryptographic community prepares for the post-quantum era, the SIKE experience reinforces the importance of maintaining multiple cryptographic approaches and the infrastructure to support transitions between them. This flexibility will be essential as the field continues to evolve and new cryptanalytic techniques emerge. The lessons from SIKE will inform not only the development of new cryptographic standards but also the strategies for deploying and maintaining them in real-world systems.

4.6 OTHER EMERGING TECHNIQUES

4.6.1 Symmetric-Key Enhancements

Symmetric-key cryptography remains the cornerstone of secure communications in the digital age, providing the fundamental mechanisms for encrypting data at rest and in transit. While asymmetric cryptography often garners more attention due to its role in key exchange and digital signatures, symmetric algorithms form the workhorse of practical encryption systems, securing everything from financial transactions to personal communications. As computational power increases and new threats emerge, the cryptographic community continues to develop innovative enhancements to symmetric-key algorithms that improve security, performance, and functionality.

The ongoing evolution of symmetric-key cryptography demonstrates the field's remarkable adaptability in responding to emerging threats and technological advancements. From advanced key scheduling to quantum-resistant designs, these enhancements ensure that symmetric encryption remains at the forefront of secure communications technology. As computational capabilities continue to advance and new attack vectors emerge, the cryptographic community's commitment to refining symmetric algorithms guarantees their continued relevance in the security landscape.

The future of symmetric cryptography lies in the balanced development of security, performance, and functionality enhancements. Innovations that address specific implementation challenges while maintaining strong theoretical security will drive the next generation of symmetric algorithms. As these techniques mature and new requirements emerge, symmetric-key cryptography will continue to evolve, providing the essential security foundations for digital communications in an increasingly complex technological environment. The remarkable progress in symmetric cryptography ensures that it will remain the workhorse of secure communications for decades to come, adapting to meet whatever challenges the future may bring.

Evolution

The development of symmetric-key cryptography has followed a remarkable trajectory from simple substitution ciphers to sophisticated modern algorithms. Early symmetric systems like the Data Encryption Standard (DES) demonstrated the power of block ciphers but eventually succumbed to brute-force attacks as computing power advanced. The introduction of the Advanced Encryption Standard (AES) marked a significant leap forward, providing stronger security through larger key sizes and more complex transformations. This evolution continues today with enhancements that address specific vulnerabilities, improve performance, and extend functionality to meet contemporary security requirements. Modern symmetric-key algorithms build upon this foundation by incorporating mathematical innovations that enhance security while optimizing computational efficiency.

Techniques such as improved diffusion and confusion properties, more sophisticated key scheduling, and advanced mode operations have transformed symmetric cryptography into a highly adaptable security tool. These enhancements reflect the cryptographic community's response to emerging threats, including side-channel attacks, implementation vulnerabilities, and the potential future impact of quantum computing. The ongoing refinement of symmetric algorithms demonstrates the field's commitment to maintaining robust security in an ever-changing technological landscape.

Advanced Key Scheduling Techniques

Key scheduling represents one of the most critical areas of enhancement in modern symmetric cryptography. The process by which a primary key is expanded into round keys directly impacts both security and performance characteristics of encryption algorithms. Recent innovations in key scheduling have focused on improving resistance to related-key attacks while optimizing computational efficiency. Techniques such as key whitening, where additional key material is combined with plaintext or ciphertext, have been refined to provide stronger security guarantees without significant performance penalties.

Another significant advancement involves dynamic key scheduling, where the key expansion process adapts based on input characteristics or environmental factors. This approach can provide enhanced resistance to certain cryptanalytic attacks while maintaining compatibility with existing implementations. Some modern algorithms incorporate key-dependent S-boxes or other transformations that vary based on the key material, making cryptanalysis more challenging. These techniques demonstrate how careful design of the key scheduling process can substantially enhance the overall security of symmetric encryption schemes.

Enhanced Block Cipher Modes of Operation

The security and functionality of symmetric encryption often depend as much on the mode of operation as on the underlying block cipher itself. Recent years have seen significant innovations in operational modes that extend the capabilities of symmetric encryption while addressing specific security concerns. Authenticated encryption modes, which combine confidentiality and integrity protection in a single primitive, represent one of the most important advancements in this area. These modes eliminate the need for separate encryption and authentication mechanisms, reducing complexity while improving security.

New modes of operation have been developed to address specific use cases and threat models. For instance, modes optimized for disk encryption provide efficient sector-by-sector processing while maintaining strong security guarantees. Other innovations include modes designed for parallel processing environments, enabling efficient encryption in modern multi-core architectures.

The development of online encryption modes that can process data streams without requiring the entire message to be available at once has expanded the applicability of symmetric encryption to new domains. These enhanced modes demonstrate how thoughtful design of operational parameters can significantly extend the practical utility of symmetric cryptography.

Improved Resistance to Implementation Attacks

As physical attacks against cryptographic implementations have become more sophisticated, symmetric-key algorithms have evolved to incorporate defenses against these threats. Side-channel resistance has become a primary focus of modern symmetric cipher design, with techniques developed to thwart timing attacks, power analysis, and electromagnetic leakage. Constant-time implementations that execute operations in uniform time regardless of secret values have become standard practice. Other innovations include masking techniques that obscure intermediate values during computation and shuffling operations to prevent leakage through memory access patterns.

Hardware-oriented enhancements have also emerged to address implementation vulnerabilities. Techniques such as bit-slicing, where operations are performed on multiple data bits simultaneously, can help obscure power consumption patterns. The development of white-box cryptography, where implementations are designed to resist reverse engineering even when attackers have full access to the execution environment, represents another significant advancement. These implementation-focused enhancements demonstrate how modern symmetric cryptography must consider not just mathematical security but also resistance to physical attacks in real-world deployments.

Quantum-Resistant Symmetric Cryptography

While quantum computing poses a more immediate threat to asymmetric cryptography, symmetric algorithms also require consideration in the post-quantum security landscape. Grover's algorithm demonstrates that quantum computers could potentially halve the effective security of symmetric keys, necessitating adjustments to key lengths. Modern symmetric algorithms are being designed with quantum resistance in mind, incorporating larger key sizes and enhanced diffusion properties to maintain security margins against potential quantum attacks.

Some emerging symmetric constructions are specifically optimized for quantum-resistant applications. These include algorithms with particularly strong diffusion characteristics that make them more resistant to quantum search techniques. The development of symmetric primitives that can be efficiently implemented in quantum computing environments represents another area of innovation. Research into symmetric algorithms that leverage quantum properties for enhanced security while remaining resistant to quantum attacks shows particular promise. These quantum-considerate enhancements ensure that symmetric cryptography remains viable in the emerging quantum computing era.

Lightweight Cryptography for Constrained Environments

The proliferation of Internet of Things (IoT) devices and embedded systems has driven demand for symmetric algorithms optimized for resource-constrained environments. Lightweight cryptography represents a specialized branch of symmetric-key enhancements focused on minimizing computational requirements while maintaining strong security guarantees. These algorithms employ simplified rounds, optimized state sizes, and efficient key schedules to reduce memory and processing requirements.

Recent innovations in lightweight cryptography include the development of algorithms specifically designed for hardware implementation, with features that map efficiently to silicon. Some constructions use bit-oriented operations rather than byte-oriented transformations to reduce circuit complexity. The creation of symmetric primitives that can be implemented with minimal gate counts while providing adequate security demonstrates the possibilities in this domain. These lightweight enhancements ensure that symmetric cryptography can be deployed in environments where traditional algorithms would be impractical, extending secure communications to the smallest and most constrained devices.

Enhanced Authentication Mechanisms

Modern symmetric cryptography has seen significant advancements in authentication capabilities beyond traditional message authentication codes. The development of algorithms that provide both encryption and authentication in a unified operation represents a major enhancement in symmetric security. These constructions often employ specialized modes of operation that interleave encryption and authentication steps to provide comprehensive protection with minimal overhead.

Some recent symmetric algorithms incorporate advanced authentication tags that provide stronger security guarantees than traditional approaches. Techniques such as polynomial-based authentication and keyed permutation-based authentication have been developed to offer enhanced resistance to forgery attempts. The creation of symmetric primitives that can verify authentication tags more efficiently while maintaining high security levels demonstrates ongoing innovation in this area. These authentication enhancements extend the capabilities of symmetric cryptography to provide more comprehensive security solutions in a single, efficient package.

Adaptive Security Techniques

Modern symmetric cryptography has incorporated adaptive techniques that modify behavior based on environmental factors or threat assessments. These enhancements include algorithms that can adjust their security parameters dynamically in response to detected threats or changing operational conditions. Some constructions employ runtime reconfiguration of cryptographic parameters to respond to potential attacks while maintaining performance.

The development of symmetric algorithms with built-in integrity verification mechanisms represents another adaptive enhancement. These techniques can detect and respond to tampering attempts or implementation anomalies without requiring external verification. The creation of self-testing symmetric primitives that can verify their own correct operation demonstrates the possibilities in this domain. These adaptive techniques ensure that symmetric cryptography can maintain security even in dynamic or potentially compromised environments.

4.6.2 Hybrid Approaches

In the modern digital landscape, the protection of data and communications has become more critical than ever before. As threats evolve and computational power increases, the need for cryptographic methods that are both efficient and secure becomes imperative. Hybrid cryptographic approaches have emerged as a powerful response to this challenge. These techniques combine the strengths of different cryptographic paradigms to create systems that are more robust, flexible, and future-proof. Hybrid cryptography is not a single algorithm or protocol but a design philosophy that integrates multiple cryptographic schemes, such as symmetric-key and asymmetric-key encryption, classical and post-quantum systems, or multiple post-quantum approaches, into unified solutions.

Hybrid cryptographic approaches embody the principle of layered defense, combining different algorithms to achieve stronger and more adaptable security. As digital ecosystems become more complex and quantum computing looms as a potential disruptor, hybrid cryptography provides a path forward that is both cautious and innovative. Whether in secure messaging, digital signatures, blockchain, or embedded systems, the application of hybrid cryptography is expanding rapidly. While challenges in integration and performance must be addressed, the benefits of enhanced security and long-term resilience make hybrid methods a critical component of modern and future cryptographic infrastructures.Hybrid cryptography represents a pragmatic and strategic approach to cryptographic evolution in the face of uncertainty. As the full capabilities of quantum computers remain uncertain in terms of timeline and scalability, hybrid methods provide a safety net that bridges the gap between classical and post-quantum cryptography. The ongoing standardization of post-quantum algorithms by organizations such as NIST and the growing support from industry giants suggest that hybrid solutions will play a central role in the cryptographic landscape for years to come. Future hybrid systems may even integrate multiple post-quantum schemes to diversify assumptions and reduce reliance on a single cryptographic hard problem.

Foundation

The basic idea behind hybrid cryptography is to leverage the advantages of each type of cryptographic method while mitigating their respective weaknesses. In traditional applications, hybrid schemes commonly pair asymmetric-key cryptography for secure key exchange with symmetric-key cryptography for actual data encryption. Asymmetric algorithms like RSA or ECC are excellent for establishing secure channels, but they are computationally expensive for large-scale data encryption. Symmetric algorithms like AES, on the other hand, are highly efficient but require a shared secret key. Hybrid systems use the former to exchange keys securely and the latter to encrypt data efficiently, achieving a balanced combination of security and performance.

Need

With the advent of quantum computing, the urgency to transition towards quantum-resistant cryptographic schemes has intensified. Quantum algorithms like Shor's algorithm threaten the security of most widely-used public-key systems. However, fully replacing classical cryptography with post-quantum alternatives is not a simple task. Many post-quantum algorithms are still under evaluation and can be computationally demanding or untested in real-world scenarios. Hybrid approaches offer a practical solution to this dilemma by combining traditional public-key schemes with post-quantum algorithms in a layered manner. This dual-encryption method ensures that even if one component becomes vulnerable, the overall system retains its security through the other component.

Post-Quantum and Classical Hybrid Encryption

One of the most active areas of research in hybrid cryptography is the development of encryption schemes that integrate classical and post-quantum methods. For instance, a message can be encrypted with a symmetric key, and this key can be encrypted using both a traditional public-key algorithm like RSA and a post-quantum algorithm like Kyber. During decryption, the receiver uses both private keys to retrieve the symmetric key and then decrypt the message. This dual-encryption approach guarantees that the message remains secure unless both the classical and quantum-resistant methods are compromised. The National Institute of Standards and Technology (NIST) has encouraged such hybrid implementations during the transition period to post-quantum cryptography, ensuring backward compatibility and increased confidence in security.

Hybrid Key Exchange Protocols

Hybrid key exchange protocols are also gaining momentum, especially in secure communication systems such as TLS (Transport Layer Security). In hybrid key exchange, the client and server establish a shared secret using a combination of classical and post-quantum key exchange algorithms. The final session key is derived from both key materials, ensuring that a compromise of one algorithm does not expose the session key.

This layered security model is particularly attractive in environments where future-proofing data confidentiality is essential. Several prototypes and research implementations have already shown that hybrid key exchange can be integrated into existing protocols with acceptable performance overhead, making it a promising candidate for deployment in critical infrastructure.

Applications in Secure Messaging

Secure messaging platforms like Signal and WhatsApp have revolutionized private communication, and hybrid cryptography is playing an increasingly important role in these environments. The double ratchet algorithm used in Signal, for example, is a hybrid construction that combines the asymmetric Diffie-Hellman exchange with symmetric-key ratcheting. This method allows the encryption keys to evolve over time, providing forward secrecy and post-compromise security. As these platforms look towards quantum-resistant alternatives, hybrid designs can integrate post-quantum key exchanges into the existing architecture without disrupting usability or performance. This hybrid adaptability is crucial for maintaining both long-term and short-term security in messaging systems.

Digital Signatures and Hybrid Authentication

In the realm of digital signatures, hybrid techniques are also being explored to enhance trust and security. A hybrid digital signature scheme may involve signing a message using both a classical algorithm such as ECDSA and a post-quantum algorithm like Dilithium or SPHINCS+. The verifier checks both signatures before accepting the authenticity of the message. This model ensures that the signature remains valid even if one of the algorithms is broken in the future. It is especially useful for applications where long-term verification is required, such as legal documents, blockchain transactions, or firmware updates. The redundancy in hybrid signatures provides assurance that the integrity and origin of a message remain verifiable despite evolving threats.

Integration Challenges and Performance Considerations

While hybrid cryptographic systems offer enhanced security, they also introduce complexity in terms of implementation and performance. Managing multiple algorithms increases the size of keys, ciphertexts, and signatures, which can affect transmission and storage efficiency. Additionally, the computational overhead of running two encryption or signing processes must be balanced against the performance requirements of the application. Careful protocol design and optimization are necessary to ensure that hybrid systems do not become a bottleneck. Standardization efforts must also address issues such as how to combine the outputs of different schemes securely, how to handle failure modes, and how to update existing infrastructure to accommodate hybrid systems without introducing vulnerabilities.

Hybrid Systems in Blockchain and Decentralized Technologies

Blockchain and decentralized systems are increasingly adopting hybrid cryptographic models to secure transactions and smart contracts. Given the irreversible nature of blockchain entries, post-quantum security is critical for ensuring the long-term integrity of records. Hybrid digital signatures can be used in blockchain to protect transactions against both current and future threats. Some research projects are exploring the integration of lattice-based post-quantum algorithms with existing elliptic curve-based methods in blockchain networks. This dual protection model is especially relevant for applications such as digital identity, cryptocurrency, and decentralized finance, where the potential value of compromised data is extremely high and long-term verification is necessary.

Hybrid Cryptography for IoT and Embedded Systems

Internet of Things (IoT) devices present unique challenges for cryptography due to constraints in memory, processing power, and energy consumption. Hybrid cryptographic approaches for IoT need to balance post-quantum resilience with lightweight operation. Researchers are working on hybrid schemes that pair lightweight symmetric ciphers with efficient post-quantum key encapsulation mechanisms, allowing secure communication between constrained devices and central servers. Some strategies involve offloading the heavy computation to the cloud while retaining lightweight cryptographic primitives on the device. These hybrid models enable IoT systems to achieve a high level of security without exceeding their hardware limitations.

CHAPTER

5 Standardization and Governance

5.1 NIST POST-QUANTUM CRYPTOGRAPHY STANDARDIZATION PROJECT

5.1.1 Process

The National Institute of Standards and Technology's Post-Quantum Cryptography (PQC) Standardization Project represents one of the most significant initiatives in modern cryptographic history. Launched in response to the growing threat posed by quantum computing to classical cryptographic systems, this multi-year effort aims to identify and standardize quantum-resistant cryptographic algorithms that can secure communications in the post-quantum era.

As quantum computers advance toward practical realization, the need for cryptographic algorithms that can withstand their computational power has become increasingly urgent. The NIST PQC project serves as a coordinated global effort to evaluate, refine, and ultimately standardize cryptographic primitives that will form the foundation of secure communications for decades to come.

The standardization process reflects NIST's long-standing role in establishing cryptographic standards that underpin global security infrastructure. From the Data Encryption Standard (DES) to the Advanced Encryption Standard (AES) and the Secure Hash Algorithm (SHA) family, NIST's standardization efforts have shaped the cryptographic landscape for generations. The PQC project continues this tradition by addressing what may be the most significant cryptographic transition since the advent of public-key cryptography. Through a rigorous, transparent, and collaborative evaluation process, NIST seeks to identify cryptographic algorithms that offer both strong security guarantees and practical implementation characteristics.

The NIST Post-Quantum Cryptography Standardization Project represents a landmark effort in cryptographic history, guiding the global transition to quantum-resistant security. Through its comprehensive and collaborative evaluation process, the project has identified cryptographic algorithms that will form the foundation of secure communications in the quantum era. The standardization of these algorithms marks both an endpoint and a beginning—the culmination of years of cryptographic research and the start of a new era in secure communications.

The process has demonstrated the complexity and importance of cryptographic transitions, highlighting the need for coordinated global efforts to address emerging threats. As the cryptographic community moves forward with the deployment of post-quantum algorithms, the lessons from this standardization effort will continue to resonate.

The NIST PQC project has not just produced new cryptographic standards but also established a model for how such transitions can be managed in the future. As quantum computing continues to advance, the cryptographic landscape will undoubtedly face new challenges, but the foundation laid by this standardization process ensures that secure communications can be maintained in the face of evolving computational threats.

As the initial phase of the NIST PQC project concludes with the selection of standardized algorithms, the work of transitioning to post-quantum cryptography continues. Future efforts will focus on the deployment and refinement of these standards, incorporating new cryptanalytic insights and implementation experiences. The cryptographic community will continue to evaluate the standardized algorithms, potentially leading to adjustments and additional rounds of standardization as understanding evolves.

The process has also highlighted the need for ongoing cryptographic research to develop new algorithms and improve existing ones. As quantum computing advances, the cryptographic landscape will continue to evolve, requiring new standardization efforts to address emerging threats. The NIST PQC project has established a model for how such transitions can be managed, with its collaborative and iterative approach serving as a template for future cryptographic standardization. The lessons learned from this process will inform not just post-quantum cryptography but also future cryptographic transitions as computational capabilities continue to advance.

Genesis and Structure of the Standardization Process

The NIST PQC Standardization Project began in 2016 with a call for proposals that attracted submissions from cryptographers worldwide. This initial phase demonstrated the global cryptographic community's recognition of the quantum threat and the need for coordinated action.

The project's structure was designed to be comprehensive and iterative, allowing for multiple rounds of evaluation and refinement based on cryptanalytic advances and implementation experiences. This approach reflects the complexity of the post-quantum transition, which requires not just theoretical security but also practical considerations of performance, compatibility, and implementation characteristics. The standardization process was organized into several phases, beginning with the submission and initial evaluation of candidate algorithms. This was followed by multiple rounds of more detailed analysis, with algorithms being selected, refined, or eliminated based on evolving understanding of their security and practical characteristics.

The process incorporated public feedback and independent cryptanalysis, ensuring that the evaluation benefited from the collective expertise of the global cryptographic community. This collaborative approach has been instrumental in identifying both the strengths and vulnerabilities of various post-quantum candidates, guiding the selection of algorithms that offer the best combination of security and practicality.

Evaluation Criteria and Methodology

The NIST PQC project established comprehensive evaluation criteria to assess the suitability of candidate algorithms for standardization. Security formed the primary consideration, with algorithms evaluated based on their resistance to both classical and quantum attacks. This involved not just theoretical analysis of the underlying mathematical problems but also practical cryptanalysis to identify potential vulnerabilities. The evaluation considered various attack vectors, including those specific to quantum computing environments, to ensure that selected algorithms would remain secure as computational capabilities advance.

Performance characteristics formed another critical evaluation dimension, as practical cryptographic algorithms must offer reasonable efficiency in real-world implementations. This included considerations of computational speed, memory requirements, and key sizes, which impact the feasibility of deployment in various environments. The evaluation also examined implementation characteristics, including resistance to side-channel attacks and suitability for different hardware platforms. These practical considerations ensured that the standardization process would result in algorithms that could be effectively deployed across diverse computing environments, from high-performance servers to resource-constrained IoT devices.

The Submission and Initial Evaluation Phase

The initial call for proposals in 2016 yielded 82 submissions from cryptographers around the world, demonstrating the global interest and investment in post-quantum cryptography. These submissions represented a diverse range of cryptographic approaches, including lattice-based, hash-based, code-based, multivariate, and isogeny-based constructions. The variety of mathematical foundations reflected the cryptographic community's exploration of different paths to quantum resistance, each with its own security assumptions and performance characteristics. The initial evaluation phase involved a comprehensive review of these submissions to assess their basic security properties and potential for standardization. This process resulted in the selection of 26 algorithms for further consideration in the second round of evaluation.

The diversity of approaches that advanced demonstrated the project's commitment to exploring multiple cryptographic paradigms rather than prematurely converging on a single mathematical foundation. This breadth of consideration ensured that the standardization process would benefit from the strengths of different cryptographic approaches, potentially leading to multiple standardized algorithms suited to various applications.

Subsequent Evaluation Rounds and Algorithm Refinement

The second round of evaluation subjected the remaining candidates to more rigorous analysis, incorporating both theoretical cryptanalysis and practical implementation considerations. This phase saw significant cryptanalytic progress, with researchers identifying vulnerabilities in some algorithms that led to their elimination from consideration. The process also involved refinement of the surviving candidates, with submitters adjusting parameters and improving implementations based on feedback from the cryptographic community.

This iterative refinement process highlighted the dynamic nature of cryptographic standardization, where algorithms evolve in response to new cryptanalytic insights. The third round further narrowed the field, with algorithms evaluated based on updated security analyses and implementation experiences. This phase demonstrated the value of the multi-round approach, as it allowed for the incorporation of new mathematical insights and practical considerations that emerged during the evaluation process. The surviving candidates represented those algorithms that best balanced security, performance, and implementation characteristics based on the accumulated knowledge from the cryptographic community.

Final Selection and Standardization

The final phase of the NIST PQC project resulted in the selection of algorithms for standardization, marking a significant milestone in the transition to quantum-resistant cryptography. The chosen algorithms represented those that had demonstrated the strongest security guarantees and most practical implementation characteristics through multiple rounds of evaluation. This selection process incorporated not just NIST's analysis but also extensive feedback from the global cryptographic community, ensuring that the standardized algorithms would meet the diverse needs of various applications. The final standards include algorithms from different cryptographic families, reflecting the project's recognition that different approaches may be suitable for different use cases. This diversity ensures that the post-quantum cryptographic landscape will not be dependent on a single mathematical foundation, providing resilience against potential future cryptanalytic advances. The standardization of multiple algorithms also allows for cryptographic agility, where systems can transition between different primitives as needed. This final selection represents the culmination of years of collaborative effort, providing a foundation for secure communications in the quantum era.

Impact on Global Cryptographic Infrastructure

The NIST PQC standardization process will have profound implications for global cryptographic infrastructure, guiding the transition from classical to post-quantum cryptography. The selected algorithms will form the basis for new cryptographic standards that will be implemented across various security protocols and systems. This transition represents one of the most significant changes in cryptographic practice since the adoption of public-key cryptography, requiring careful planning and coordination across industries and governments.

The standardization effort has already stimulated significant activity in the cryptographic community, with researchers and practitioners preparing for the deployment of post-quantum algorithms. This includes developing implementation guidelines, performance optimizations, and migration strategies for existing systems. The process has also highlighted the importance of cryptographic agility, where systems can adapt to new algorithms as standards evolve. The global impact of this standardization effort underscores the critical role that NIST plays in establishing cryptographic standards that underpin secure communications worldwide.

Challenges and Lessons Learned

The NIST PQC standardization process has encountered several challenges that have provided valuable lessons for the cryptographic community. One significant challenge has been the dynamic nature of cryptanalysis, where new mathematical insights can dramatically alter the security landscape. The attack on SIKE during the evaluation process demonstrated how quickly the perceived security of an algorithm can change, highlighting the need for continuous cryptanalysis even after standardization. This experience reinforced the importance of maintaining flexibility in cryptographic systems to adapt to new developments.

Another challenge has been balancing the various considerations that go into cryptographic standardization, including security, performance, and implementation characteristics. The process has shown that these factors are often interdependent, with improvements in one area potentially impacting others. The need to consider not just theoretical security but also practical deployment characteristics has been a recurring theme throughout the evaluation. These challenges have provided important lessons for future standardization efforts, emphasizing the complexity of cryptographic transitions and the need for comprehensive evaluation processes.

5.1.2 Rounds

The rounds of the NIST Post-Quantum Cryptography Standardization Project represent a critical turning point in the history of cryptography. Through a transparent, collaborative, and rigorous process, NIST has guided the global community towards cryptographic algorithms that can withstand the power of quantum computation. As the project progresses into future rounds and adoption accelerates, it will shape the next generation of secure digital infrastructure. The standardized post-quantum algorithms emerging from this project will serve as the foundation for secure communication, data integrity, and digital trust in the decades to come. The proactive and careful planning of NIST and its partners ensures that the transition to quantum-safe cryptography will be both secure and efficient, safeguarding global digital ecosystems against the quantum threat.

First Round: Call for Proposals and Initial Submissions

The first round of the NIST PQC project began with the call for submissions, which concluded in November 2017. A total of sixty-nine algorithm candidates were accepted for evaluation. These submissions represented a diverse set of mathematical foundations, including lattice-based, code-based, multivariate polynomial, hash-based, and supersingular isogeny-based schemes. NIST and the broader cryptographic community began the process of evaluating these algorithms based on their security, performance, and implementability on various platforms. The first round was a broad survey of potential solutions and aimed to filter out candidates with apparent weaknesses or impractical designs.

Second Round: Focused Evaluation and Refinement

In January 2019, NIST announced the beginning of the second round with a narrowed pool of twenty-six candidate algorithms. This group included seventeen encryption/KEM schemes and nine digital signature algorithms. During this phase, further cryptanalysis, implementation testing, and optimization analysis were carried out. Submitters were also allowed to revise and resubmit their proposals to address concerns or improve efficiency. The second round provided deeper insight into the strengths and weaknesses of each candidate, and it was during this stage that many promising algorithms began to distinguish themselves. However, a few candidates were withdrawn due to emerging cryptanalytic attacks or performance limitations.

Third Round: Finalists and Alternates Announced

The third round of the NIST PQC Standardization Project commenced in July 2020 with the selection of seven finalist algorithms and eight alternate candidates. The finalists for public-key encryption and KEMs included algorithms such as Kyber, NTRU, and Saber, while signature finalists included Dilithium, Falcon, and Rainbow. These algorithms were considered the most promising candidates for standardization. The alternate candidates were retained for further analysis and as potential replacements should any weaknesses arise in the finalists. The third round emphasized final performance tuning, resistance to side-channel attacks, and suitability for constrained environments such as embedded systems and the Internet of Things (IoT).

Selection Criteria and Evaluation Standards

Throughout all rounds, NIST applied a consistent set of criteria to evaluate candidate algorithms. The most important factor was security, particularly resistance to quantum attacks. However, performance on both hardware and software platforms, key and ciphertext sizes, algorithm complexity, and implementation simplicity also played crucial roles. Algorithms that were resistant to known classical and quantum attacks and exhibited robustness against implementation vulnerabilities were prioritized. NIST also looked at the versatility of algorithms across various application domains, including digital communications, secure storage, and secure transactions.

Announcement of Standardized Algorithms

In July 2022, NIST announced the first group of selected algorithms to be standardized. For key establishment, Kyber was chosen as the primary KEM to be standardized due to its strong security, high efficiency, and manageable key sizes. For digital signatures, NIST selected three algorithms: Dilithium, Falcon, and SPHINCS+. Dilithium and Falcon are lattice-based and offer a balance between performance and compactness, while SPHINCS+ is a stateless hash-based scheme known for its conservative security assumptions and long-term reliability. This selection marked a significant milestone, signaling the beginning of the transition from traditional public-key cryptography to post-quantum secure systems.

Future Rounds and Ongoing Standardization

Even after the initial standard selections, NIST emphasized that the PQC standardization process is not complete. The project continues with additional rounds to evaluate alternate candidates and to potentially standardize more algorithms. These additional standards may address specific use cases or provide diversity in cryptographic assumptions to safeguard against unforeseen weaknesses. NIST has also initiated calls for public feedback on draft standards and implementation guidelines to ensure broad consensus and practical viability. Furthermore, hybrid schemes that combine classical and post-quantum cryptography are being studied as transitional solutions during the widespread deployment of PQC.

Implementation and Real-World Adoption

Following the selection of standardized algorithms, attention has shifted to implementation and integration. Major software libraries, open-source cryptographic toolkits, and industry vendors have started adopting the selected algorithms.

Tech companies such as Google, Microsoft, and IBM are actively testing hybrid and quantum-safe algorithms in applications ranging from secure web browsing to encrypted messaging. Challenges remain, especially in retrofitting existing systems and protocols to accommodate new key sizes, signature formats, and performance profiles. However, guidance from NIST and collaborative community efforts are helping to ease this transition.

Global Collaboration and Impact on Standards

The NIST PQC Standardization Project has not only impacted U.S. cryptographic standards but has also influenced global practices. Organizations such as ISO, ETSI, and IETF have closely followed NIST's process and are incorporating PQC recommendations into their respective frameworks. The international nature of the project has helped build confidence in the selected algorithms and ensured that they meet diverse requirements. By fostering cooperation across countries, academia, and industry, NIST's PQC project is setting a global benchmark for secure cryptographic standards in the quantum era.

5.1.3 Finalists

NIST's Post-Quantum Cryptography Standardization Project finalists represent the culmination of years of rigorous scientific evaluation and represent the foundation for quantum-resistant security in the coming decades. The selected algorithms—CRYSTALS-Kyber, CRYSTALS-Dilithium, Falcon, and SPHINCS+—offer a balanced portfolio addressing different security requirements and implementation constraints. Their standardization marks the beginning of a complex but necessary transition that will reshape cryptographic practice across industries and applications.

While challenges remain in implementation, performance optimization, and ecosystem adoption, NIST's thorough process provides confidence that these algorithms can protect sensitive data against the quantum threat. As organizations begin migrating to these new standards, continued cryptanalysis, implementation experience, and protocol integration will further refine their deployment, ensuring a secure digital infrastructure ready for the quantum computing era. The project's success demonstrates how proactive, collaborative standardization can address emerging technological threats before they materialize.

Lattice-Based Finalists: CRYSTALS-Kyber and CRYSTALS-Dilithium

Lattice-based cryptography emerged as the dominant approach among NIST's finalists, with CRYSTALS-Kyber selected as the primary Key Encapsulation Mechanism (KEM) and CRYSTALS-Dilithium as the main digital signature scheme. Kyber leverages the hardness of Module-Learning With Errors (MLWE), a variant of the Learning With Errors (LWE) problem, to provide efficient key exchange with strong security guarantees. Its compact key sizes and fast operations make it suitable for a wide range of applications, from IoT devices to high-speed network encryption. Dilithium, also based on structured lattices, offers similarly attractive properties for digital signatures, balancing security with practical performance. Both algorithms benefit from conservative security reductions to well-studied lattice problems and demonstrate resistance to known quantum and classical attacks. Their algebraic structure enables relatively straightforward implementation while maintaining flexibility for optimization across different hardware platforms.

Alternate Finalists: Falcon and SPHINCS+

To ensure diversity in cryptographic approaches, NIST designated additional finalists as alternatives to the primary lattice-based selections. Falcon, a lattice-based signature scheme using NTRU lattices, was chosen for applications requiring particularly compact signatures, such as blockchain systems and embedded devices. While Falcon offers smaller signature sizes than Dilithium, its implementation complexity—particularly the need for floating-point arithmetic and careful side-channel protection—makes it more challenging to deploy securely. For scenarios demanding hash-based security, SPHINCS+ represents a conservative choice as a stateless signature scheme. Relying solely on the security of cryptographic hash functions, SPHINCS+ provides strong security guarantees but comes with larger signature sizes and higher computational costs compared to lattice-based alternatives. These alternate finalists give implementers options when the primary choices may not suit specific use cases or security requirements.

Security Considerations and Parameter Selection

The NIST finalists were selected only after extensive cryptanalysis by the global research community. Each algorithm's security parameters were carefully calibrated to meet NIST's defined security categories: Level 1 (comparable to AES-128), Level 3 (AES-192), and Level 5 (AES-256). The lattice-based schemes derive their security from the worst-case hardness of lattice problems, which remain resistant to both classical and quantum attacks. SPHINCS+'s security stems directly from the collision resistance of the underlying hash function, providing a hedge against potential future breakthroughs in algebraic cryptanalysis. NIST's parameter choices reflect a balance between conservative security margins and practical performance, with options for increased parameters if future cryptanalysis reveals weaknesses. The evaluation process specifically considered resistance to known quantum algorithmic approaches, ensuring the finalists maintain their security claims even against large-scale quantum computers.

Performance Characteristics and Implementation Challenges

Practical deployment considerations played a significant role in NIST's final selections. The lattice-based finalists generally offer excellent performance across a range of platforms, with Kyber demonstrating encryption/decryption speeds comparable to current elliptic curve cryptography. Dilithium's signature generation and verification times are practical for most applications, though noticeably slower than current ECDSA signatures.

Falcon's signature generation is computationally intensive but produces signatures nearly an order of magnitude smaller than Dilithium's. SPHINCS+, while having the strongest security foundations, suffers from large signature sizes (ranging from 8KB to 30KB depending on parameters) and slower operations. Implementation challenges vary across the finalists, with lattice-based schemes requiring careful attention to constant-time arithmetic to prevent timing side-channels, and Falcon demanding high-precision floating-point implementations. These performance trade-offs allow organizations to select algorithms based on their specific constraints regarding computation, bandwidth, and storage.

Standardization Process and Timeline

NIST's standardization approach has been methodical and transparent, involving multiple rounds of public scrutiny over six years. The third round (2019-2022) saw intensive cryptanalysis of candidate algorithms, with several promising schemes being broken or showing vulnerabilities that led to their elimination. The final announcement in July 2022 established a clear migration path, with draft standards expected in 2023-2024. NIST plans to maintain these standards for at least two decades, with provisions for periodic review and potential updates as cryptanalysis advances. The standardization process includes detailed specification documents, reference implementations, and test vectors to ensure interoperability. This thorough approach aims to facilitate smooth adoption across industries while maintaining flexibility to incorporate future improvements or respond to emerging cryptographic threats.

Industry Adoption and Migration Considerations

The transition to NIST's post-quantum standards presents significant challenges for industry adoption. Cryptographic libraries and protocols must be updated to support the new algorithms, requiring coordinated changes across software stacks. Hybrid deployment strategies—combining classical and post-quantum cryptography—are recommended during the transition period to maintain backward compatibility while adding quantum resistance.

Performance characteristics of the finalists will influence their adoption patterns, with Kyber likely seeing early deployment in TLS and VPN protocols, while Dilithium may become the default for document signing. Specialized applications may prefer Falcon for its compact signatures or SPHINCS+ for its conservative security properties. Organizations must inventory their cryptographic assets, prioritize systems for migration, and develop testing frameworks to validate implementations against NIST's specifications. The migration process is expected to span several years as the ecosystem matures and best practices emerge.

Future-Proofing Against Quantum Advances

NIST's finalists were selected not just for their current security, but for their ability to withstand future advances in both quantum and classical cryptanalysis. The mathematical problems underlying the lattice-based schemes have resisted decades of intensive study, suggesting long-term viability. SPHINCS+'s hash-based design provides a fallback option should unexpected weaknesses emerge in structured lattice problems.

NIST has also initiated a fourth round to evaluate additional algorithms, particularly focusing on alternative approaches like multivariate and code-based cryptography to diversify the post-quantum toolkit. This forward-looking approach ensures that the standards can evolve as the threat landscape changes, maintaining protection against increasingly sophisticated quantum attacks while accommodating new cryptographic innovations.

Global Implications and International Collaboration

The NIST PQC standardization project has become a de facto global benchmark, with international standards bodies and technology companies aligning their post-quantum strategies with NIST's finalists. The European Telecommunications Standards Institute (ETSI) and the International Organization for Standardization (ISO) are developing complementary standards based on NIST's selections. This harmonization is critical for maintaining global interoperability in secure communications. However, some nations are pursuing independent post-quantum cryptography initiatives, reflecting both technical sovereignty concerns and different risk assessments about quantum threats. The widespread adoption of NIST's standards will depend on continued international cooperation, transparent evaluation processes, and shared confidence in the selected algorithms' long-term security.

5.1.4 Alternatives

NIST received hundreds of candidate algorithms, reflecting a vibrant global effort to develop quantum-resistant cryptography. Through a careful selection process, promising algorithms are shortlisted for further analysis and testing, with the goal of establishing a set of standards that can be adopted worldwide. The project emphasizes not only the security of these algorithms against quantum attacks but also their practicality in real-world applications, including internet security protocols, data encryption, digital signatures, and key exchange mechanisms.

The importance of this initiative is underscored by the widespread reliance on cryptographic systems for securing financial transactions, government communications, private data, and critical infrastructure. As quantum computers continue to evolve, the urgency to transition to quantum-resistant algorithms becomes increasingly apparent, and NIST's efforts aim to provide the authoritative guidelines needed for this transition. The standardization process is transparent and collaborative, involving input from academia, industry, and government agencies, ensuring that the selected algorithms are robust, versatile, and adaptable to various technological environments.

Furthermore, the project takes into account the need for algorithms to be efficient enough for deployment on diverse hardware, including constrained devices such as IoT sensors and mobile devices, which are often limited in computational power and bandwidth. As the NIST process progresses, some candidate algorithms have shown promising results in balancing security and performance, while others have been eliminated or deferred for further analysis. The outcome of this initiative will not only influence future cryptographic standards but also shape global cybersecurity policies, as organizations begin to adopt these new algorithms to protect sensitive data against future quantum threats.

The transition to post-quantum cryptography will require considerable effort in updating existing systems, developing new protocols, and ensuring interoperability across different platforms. NIST's work provides a critical foundation for this shift, offering confidence that the selected algorithms will form the backbone of secure communications in the quantum era. In addition to standardization, the project encourages ongoing research and innovation, recognizing that the field of quantum-resistant cryptography is dynamic and evolving. As quantum computing hardware continues to improve, so too must the cryptographic techniques designed to counteract potential threats.

The global community's collaborative approach to this project also highlights the importance of international cooperation in establishing secure standards that transcend national boundaries. Ultimately, the NIST Post-Quantum Cryptography Standardization Project plays a vital role in safeguarding the future of digital security. By proactively identifying and endorsing algorithms resistant to quantum attacks, it aims to ensure that the confidentiality, integrity, and authenticity of digital information remain intact in the face of rapidly advancing quantum technology.

This initiative exemplifies a forward-looking strategy that balances scientific innovation, practical implementation, and policy development to prepare the world for the quantum age. The success of this project will depend on continued research, broad acceptance, and seamless integration of new cryptographic standards into existing and future systems, securing the digital landscape against threats that are yet to fully materialize but are already on the horizon.

5.2 ISO/IEC AND INTERNATIONAL STANDARDIZATION EFFORTS

The development and adoption of cryptographic standards represent a cornerstone of modern information security, providing the foundation for secure communications and data protection across global systems. Within this landscape, the International Organization for Standardization (ISO) and the International

Electrotechnical Commission (IEC) play pivotal roles through their joint technical committee, ISO/IEC JTC 1/SC 27, which focuses on information security, cybersecurity, and privacy protection. These international standardization efforts ensure that cryptographic algorithms and protocols meet rigorous security requirements while maintaining interoperability across diverse systems and jurisdictions. As cybersecurity threats evolve and new technologies emerge, the work of ISO/IEC in establishing cryptographic standards becomes increasingly vital for maintaining global security infrastructure.

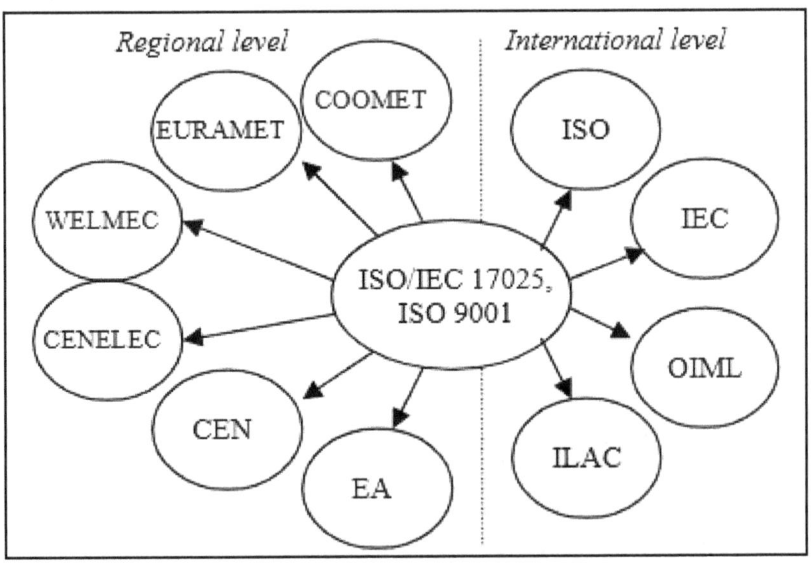

Fig: Implementation International Standards

The work of ISO/IEC in developing international cryptographic standards plays a crucial role in maintaining global security infrastructure. Through collaborative processes that incorporate diverse expertise and perspectives, the organization creates standards that provide robust security while remaining practical for implementation. The rigorous evaluation and refinement processes ensure that resulting standards can withstand evolving threats while meeting the needs of various technological environments. As cybersecurity challenges grow more complex and computational capabilities advance, the importance of coordinated international standardization efforts becomes ever more apparent.

The development of post-quantum cryptographic standards exemplifies the vital role that ISO/IEC plays in preparing for future security requirements. By facilitating global consensus on cryptographic best practices, the organization helps ensure that secure communications can be maintained across international boundaries and technological platforms.

The challenges faced in cryptographic standardization underscore the complexity of creating globally applicable security solutions, while the successes demonstrate the value of collaborative international efforts. As digital systems continue to evolve, the work of ISO/IEC in developing and maintaining cryptographic standards will remain essential for protecting information and communications in an interconnected world.

As cryptographic technologies continue to evolve, ISO/IEC standardization efforts must adapt to address emerging security requirements. The increasing complexity of cyber threats and the development of new computational paradigms necessitate continuous updates to cryptographic standards. Future standardization efforts will likely focus on areas such as homomorphic encryption, which enables computation on encrypted data, and advanced authentication mechanisms that provide stronger security guarantees. The organization will also need to address the security requirements of emerging technologies like quantum communications and advanced cryptographic primitives.

The development of standards for cryptographic agility represents another important future direction. As the transition to post-quantum cryptography demonstrates, the ability to adapt to new algorithms and security requirements will be crucial for maintaining long-term security. Future standards may incorporate mechanisms for algorithm agility, where systems can switch between different cryptographic primitives as needed. The standardization process will also need to consider the security requirements of increasingly interconnected systems, where cryptographic protection must extend across complex networks of devices and services. These future directions highlight the ongoing importance of cryptographic standardization in maintaining secure digital infrastructure.

5.2.1 Structure of ISO/IEC Cryptographic Standardization

ISO/IEC cryptographic standardization operates through a well-defined structure that facilitates international collaboration while maintaining technical rigor. The joint technical committee JTC 1/SC 27 serves as the primary body responsible for information security standards, with working groups dedicated to specific areas such as cryptographic techniques, security management, and identity management. This organizational framework allows for focused development of standards while ensuring coordination across related security domains. The standardization process typically follows several stages, beginning with the proposal of new work items and progressing through committee drafts, public review periods, and final approval as international standards.

The development of cryptographic standards within ISO/IEC involves multiple layers of review and consensus-building. National standards bodies participate in the process, providing input that reflects regional requirements and perspectives. This multi-stakeholder approach helps ensure that resulting standards address global needs while remaining practical for implementation across different technological environments. The process also incorporates liaison relationships with other standards organizations, creating a network of coordinated efforts that prevents duplication and promotes harmonization of security requirements. This structured approach to standardization enables the creation of robust cryptographic standards that can be widely adopted while maintaining flexibility for specific implementation needs.

5.2.2 Key Cryptographic Standards and their Evolution

ISO/IEC has developed numerous cryptographic standards that form the backbone of modern security infrastructure. The ISO/IEC 18033 series, for instance, provides specifications for encryption algorithms, hash functions, digital signatures, and other cryptographic primitives. These standards evolve through periodic reviews that incorporate new cryptanalytic insights and technological developments. The ISO/IEC 19790 series addresses security techniques for authentication frameworks, while ISO/IEC 29192 focuses on lightweight cryptography for constrained environments. Each of these standards undergoes rigorous evaluation to ensure they meet contemporary security requirements while remaining practical for implementation. The evolution of these standards reflects the changing landscape of cryptographic security. As new threats emerge and computational capabilities advance, ISO/IEC standards must adapt to maintain their effectiveness. This evolution often involves the incorporation of new cryptographic primitives, adjustments to parameter sizes, and the development of enhanced security mechanisms. The standardization process also considers the need for backward compatibility, ensuring that updates to cryptographic standards can be implemented without disrupting existing systems. This balance between innovation and stability characterizes the development of ISO/IEC cryptographic standards, which must serve both current security needs and future requirements.

5.2.3 Role of International Collaboration

International collaboration forms the foundation of ISO/IEC standardization efforts, bringing together diverse expertise to create globally applicable cryptographic standards. The participation of national standards bodies from various countries ensures that standards address a wide range of security requirements and implementation environments. This collaborative approach helps identify potential conflicts between regional regulations and global standards, facilitating the development of solutions that can be widely adopted. The process also benefits from contributions by industry experts, academic researchers, and government representatives, creating a comprehensive perspective on security needs.

The collaborative nature of ISO/IEC standardization extends to coordination with other international standards organizations. Relationships with bodies such as the International Telecommunication Union (ITU) and the Internet Engineering Task Force (IETF) help ensure that cryptographic standards align with related technical specifications. This inter-organizational cooperation prevents duplication of efforts and promotes the development of complementary standards that work together to provide comprehensive security solutions. The resulting standards reflect a global consensus on cryptographic best practices, enabling secure communications and data protection across international boundaries.

5.2.4 Post-Quantum Cryptography Standardization Efforts

As the cryptographic community prepares for the transition to quantum-resistant algorithms, ISO/IEC has become increasingly involved in post-quantum cryptography standardization. The organization works in coordination with other standards bodies to develop guidelines for the adoption of post-quantum cryptographic algorithms that can withstand attacks from quantum computers. This effort involves evaluating candidate algorithms from various mathematical families, including lattice-based, hash-based, code-based, and multivariate constructions. The standardization process considers not only the security of these algorithms but also their performance characteristics and implementation requirements.

The development of post-quantum cryptographic standards presents unique challenges that require international coordination. Different regions may have varying timelines for quantum computing development and different risk assessments regarding the quantum threat. ISO/IEC standardization efforts help harmonize these perspectives, creating a shared understanding of post-quantum security requirements. The process also addresses the need for cryptographic agility, where systems can transition between different algorithms as standards evolve. This international coordination ensures that the global community can maintain secure communications as quantum computing capabilities advance, preventing fragmentation of security standards that could create vulnerabilities.

5.2.5 Challenges in International Cryptographic Standardization

The development of international cryptographic standards faces several significant challenges that require careful navigation. One primary challenge involves balancing the need for strong security guarantees with practical implementation considerations. Standards must be sufficiently robust to protect against current and emerging threats while remaining feasible for deployment across various technological environments. This often requires making trade-offs between security strength, performance characteristics, and compatibility requirements. The standardization process must also consider the varying capabilities of different systems, from high-performance servers to resource-constrained IoT devices.

Another challenge arises from the diverse regulatory environments and security requirements across different jurisdictions. National security concerns, export controls, and regional privacy regulations can create complexities in developing globally applicable standards. The standardization process must navigate these considerations while maintaining the integrity of the technical specifications. Intellectual property concerns also present challenges, as standards must be developed in ways that allow for broad implementation without infringing on proprietary technologies. These challenges require careful coordination and consensus-building among the international community of experts participating in the standardization process.

5.2.6 Standardization Process and Public Review

The ISO/IEC standardization process incorporates multiple stages of review and refinement to ensure the quality and applicability of cryptographic standards. Initial proposals undergo evaluation by technical committees, which assess their suitability for standardization based on security properties, implementation characteristics, and alignment with existing standards. Successful proposals then progress to committee draft stages, where they are refined through iterative review processes. Public review periods provide opportunities for broader community input, allowing industry practitioners and academic researchers to identify potential issues and suggest improvements.

This multi-stage review process helps ensure that resulting standards reflect both expert consensus and practical considerations. The public review component is particularly important for cryptographic standards, as it allows for identification of potential vulnerabilities or implementation challenges that might not be apparent to the initial development team. Feedback from these review periods often leads to significant refinements of proposed standards, improving their security and practicality. The final approval process incorporates all received input, resulting in standards that have undergone comprehensive evaluation and benefit from diverse perspectives.

5.2.7 Implementation and Adoption Considerations

The development of cryptographic standards extends beyond their technical specifications to include guidance on implementation and adoption. ISO/IEC standards often incorporate annexes or companion documents that provide recommendations for secure implementation practices. These guidelines address potential pitfalls in cryptographic deployment, such as side-channel vulnerabilities, improper parameter selection, or insecure key management. The standardization process also considers the need for migration strategies that allow organizations to transition from older cryptographic algorithms to newer standards without disrupting operations.

The adoption of cryptographic standards requires coordination across various stakeholders, including technology vendors, system integrators, and end users. ISO/IEC works to facilitate this adoption through educational initiatives, implementation profiles, and conformity assessment programs. These efforts help ensure that standards are not only technically sound but also practically implementable across different technological environments. The organization also considers the need for interoperability testing, which verifies that different implementations of the same standard can work together effectively.

The CNSA 2.0 Suite represents a strategic and timely response to the challenges posed by quantum computing. By outlining a clear path for the adoption of post-quantum cryptographic algorithms and supporting hybrid implementations, NSA has provided national security agencies with the tools needed to safeguard critical information well into the future. CNSA 2.0 is not just a technical specification but a comprehensive security framework grounded in rigorous analysis, risk management, and interagency collaboration. As quantum computing capabilities mature, the importance of robust cryptographic defenses will only grow. In this evolving landscape, CNSA 2.0 serves as a critical foundation for the next generation of secure national defense systems.

Looking ahead, CNSA 2.0 is not intended as a final endpoint but as a major step in a longer journey towards post-quantum security. NSA will continue to monitor developments in quantum computing, cryptanalysis, and cryptographic research. The agency is expected to refine and expand CNSA 2.0 as new information emerges, including the potential inclusion of additional algorithms from future

NIST competitions or community proposals. NSA's long-term vision includes the integration of quantum-resistant cryptography into the entire lifecycle of secure systems, from hardware design to protocol implementation and key management. As part of this vision, the agency advocates for the development of quantum-safe infrastructures that are both secure and adaptable.

5.3 ROLE OF NSA'S CNSA 2.0 SUITE

Cryptographic standards are the backbone of secure communication and data protection in modern digital systems. For national security agencies, the stakes are even higher, as any vulnerability in cryptographic protocols can compromise sensitive military, diplomatic, and intelligence operations. In the United States, the National Security Agency (NSA) plays a pivotal role in guiding and enforcing cryptographic standards for use within national security systems. One of the NSA's major initiatives in recent years is the Commercial National Security Algorithm (CNSA) Suite, specifically its updated version, CNSA 2.0. This suite has been introduced in anticipation of the quantum computing threat and is designed to ensure long-term protection of classified and sensitive national security information.

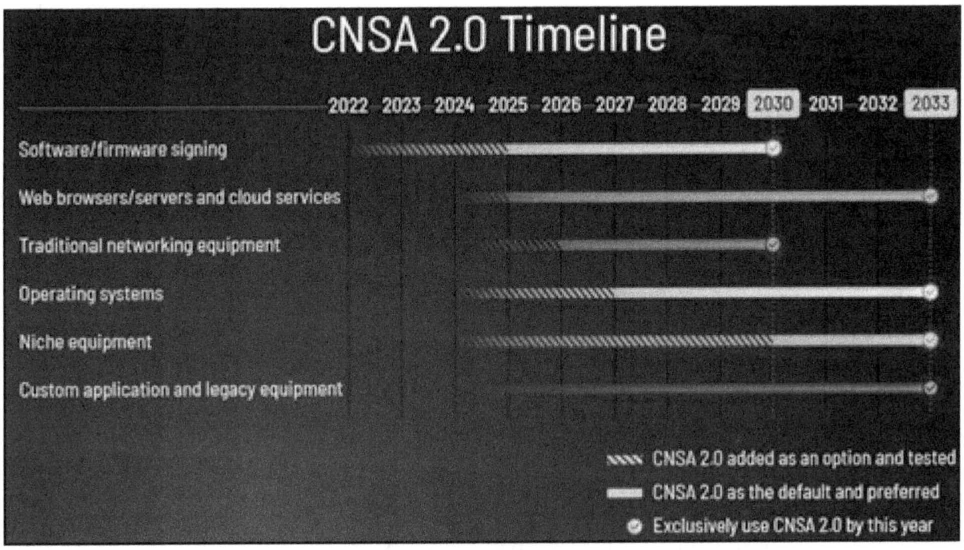

Fig: Commercial National Security Algorithm Suite

5.3.1 Evolution from Suite B to CNSA

Historically, NSA relied on Suite B cryptographic algorithms, which included well-known protocols such as AES for symmetric encryption, ECDSA for digital signatures, and ECDH for key exchange. However, the rise of quantum computing prompted a strategic shift. Suite B algorithms, which are based on traditional number-theoretic problems such as factoring and discrete logarithms, are vulnerable to quantum attacks, particularly Shor's algorithm. Recognizing this threat, NSA initiated the transition from Suite B to the CNSA Suite in 2015. The first iteration of CNSA replaced elliptic-curve algorithms with longer key length RSA and finite field Diffie-Hellman, which offered more conservative security margins, albeit without true quantum resistance. CNSA was intended as an interim measure, bridging the gap until suitable post-quantum cryptographic standards could be developed and adopted.

5.3.2 Introduction and Objectives of CNSA 2.0

CNSA 2.0 was formally announced in September 2022 as a forward-looking update to the original CNSA Suite. It represents NSA's strategic roadmap for securing national security systems against both classical and quantum adversaries. The primary objective of CNSA 2.0 is to guide government agencies and contractors in the adoption of cryptographic algorithms that will remain secure in a post-quantum world. CNSA 2.0 explicitly incorporates algorithms selected from the National Institute of Standards and Technology (NIST) Post-Quantum Cryptography Standardization Project, reflecting NSA's close collaboration with NIST and its trust in the open peer-reviewed selection process. While CNSA 2.0 does not yet mandate exclusive use of post-quantum algorithms, it establishes the foundation for phased migration and hybrid implementation strategies.

5.3.3 Incorporation of Post-Quantum Algorithms

One of the most significant features of CNSA 2.0 is its explicit recommendation of quantum-resistant algorithms, especially for key establishment and digital signatures. NSA has identified CRYSTALS-Kyber and CRYSTALS-Dilithium as foundational elements in the post-quantum transition. CRYSTALS-Kyber is a lattice-based key encapsulation mechanism, while CRYSTALS-Dilithium is a lattice-based digital signature algorithm. Both were selected as part of NIST's PQC standardization effort and have been recognized for their strong security proofs, efficient performance, and resilience against quantum and classical attacks. CNSA 2.0 provides implementation guidelines for these algorithms and encourages hybrid deployments alongside legacy algorithms to ensure backward compatibility and transitional security during migration phases.

5.3.4 Hybrid Cryptographic Implementations

To maintain interoperability and minimize risks during the transition to quantum-resistant cryptography, CNSA 2.0 supports hybrid cryptographic approaches. These involve combining classical algorithms such as RSA or ECDH with post-quantum algorithms like Kyber in a single cryptographic protocol. Hybrid models offer layered protection, ensuring that even if one algorithm becomes compromised in the future, the other layer maintains the security of the system. This is particularly important for systems that protect information requiring long-term confidentiality, such as classified military or diplomatic data. NSA's promotion of hybrid models reflects a pragmatic recognition that immediate wholesale migration to post-quantum cryptography may not be feasible across all systems.

5.3.5 Timeline and Implementation Guidance

CNSA 2.0 provides a structured timeline for adoption and encourages early testing and integration of post-quantum algorithms. NSA has recommended that national security systems begin to transition to CNSA 2.0 algorithms by 2025, with full adoption expected by the end of the decade.

This proactive timeline is designed to allow agencies and contractors sufficient time to evaluate, implement, and optimize post-quantum cryptographic solutions within their existing infrastructures. The guidance also includes the use of Federal Information Processing Standards (FIPS) validated implementations and references draft NIST Special Publications for further technical details. By aligning its roadmap with NIST's timeline for finalizing PQC standards, NSA ensures that CNSA 2.0 remains compatible with broader federal and commercial security strategies.

5.3.6 Application Domains of CNSA 2.0

CNSA 2.0 is designed specifically for national security systems, which include any systems that process classified information or are otherwise critical to national defense. This includes secure communication platforms, military command and control systems, government databases, satellite communication links, and intelligence networks. The algorithms and standards promoted in CNSA 2.0 are expected to be adopted by defense contractors, intelligence agencies, and other federal organizations responsible for handling sensitive data. In these environments, the cost of cryptographic failure is extraordinarily high, making the rigorous and forward-compatible nature of CNSA 2.0 essential to mission assurance.

5.3.7 Security Philosophy and Risk Management

NSA's approach to CNSA 2.0 reflects a conservative security philosophy that prioritizes robustness over performance. While some post-quantum algorithms may involve larger keys or slower computation compared to classical counterparts, NSA considers these trade-offs acceptable in exchange for greater long-term security guarantees. CNSA 2.0 emphasizes the importance of cryptographic agility, allowing systems to quickly adapt to new threats and updated standards. The suite also encourages regular cryptographic reviews and the implementation of update mechanisms to enable seamless replacement of algorithms if future vulnerabilities are discovered. This risk-aware mindset ensures that national security systems remain resilient in the face of both known and unknown adversarial capabilities.

5.3.8 Interagency and International Collaboration

NSA's development of CNSA 2.0 has occurred in close coordination with other U.S. government agencies, including NIST, the Department of Defense (DoD), and the Cybersecurity and Infrastructure Security Agency (CISA). By harmonizing cryptographic standards across these agencies, the U.S. government ensures unified security policies and simplifies the deployment of secure systems. Furthermore, CNSA 2.0 contributes to international discussions on quantum-safe cryptography, helping to shape global standards. The adoption of CNSA 2.0 by allies and partners strengthens collective cybersecurity resilience and facilitates secure information sharing across borders.

5.4 OPEN-SOURCE PROJECTS AND GLOBAL ECOSYSTEM

Open-source software has transformed from a niche movement into a fundamental pillar of modern technology infrastructure. What began with Richard Stallman's GNU Project and Linus Torvalds' Linux kernel has evolved into a global phenomenon shaping industries from cloud computing to artificial intelligence. The collaborative nature of open-source development has enabled unprecedented innovation cycles, where developers worldwide contribute to projects that power critical systems across all sectors. This decentralized model of software creation challenges traditional proprietary development paradigms, offering transparency, flexibility, and community-driven problem solving.

The open-source movement has matured into essential global infrastructure that transcends its software origins. Like physical infrastructure, it requires ongoing maintenance, governance, and investment despite its "invisible" nature to most end users. The ecosystem's health depends on recognizing open-source as a public good that enables innovation across all sectors of the digital economy.

Future challenges include establishing sustainable funding models, improving security practices at scale, and ensuring equitable participation across geographic and demographic boundaries. As open-source methodologies influence adjacent fields like open data, open science, and open government, their core principles of transparency, collaboration, and meritocracy offer a compelling model for solving complex global problems. The continued evolution of this ecosystem will shape not just technology, but how humanity organizes collective knowledge work in the 21st century.

The open-source ecosystem stands at several critical junctures that will define its next decade. WebAssembly (Wasm) promises to make open-source components portable across programming language boundaries, while AI-assisted coding tools like GitHub Copilot challenge traditional notions of authorship and attribution. Decentralized technologies (blockchain, federated systems) offer new models for open-source governance and value distribution. The Open Source Initiative's exploration of "ethical source" licenses reflects growing desire to enforce behavioral restrictions beyond pure software freedoms. Climate computing initiatives demonstrate open-source's potential to address global challenges through collaborative modeling and data sharing. As open-source principles extend into hardware (RISC-V), biotechnology (Open Insulin), and policy making, the ecosystem continues to prove its adaptability across domains far beyond software development.

5.4.1 Economic Impact of Open-Source Ecosystems

The global economic value generated by open-source software defies conventional measurement yet clearly represents one of the most significant productivity multipliers in modern history. Studies estimate that open-source contributes hundreds of billions annually to the global economy through direct cost savings and enabled innovation. Companies ranging from tech giants to startups now build their products on open-source foundations, with the average application incorporating over 150 open-source components.

This shared infrastructure reduces redundant development efforts while creating new business models around support, customization, and managed services. The Linux Foundation's 2022 report revealed that collaborative open-source projects generate value exceeding the GDP of many nations, demonstrating how shared digital commons can drive widespread economic growth. However, this economic model also presents challenges in sustaining non-commercial maintenance of critical infrastructure components that lack obvious monetization pathways.

5.4.2 Governance Models in Major Open-Source Projects

Successful open-source initiatives have developed sophisticated governance structures to manage their global communities while maintaining technical direction. The Apache Software Foundation's meritocratic model emphasizes earned authority through sustained contributions, while the Linux Foundation employs a membership-based approach funding critical infrastructure projects. Python's governance transition from Benevolent Dictator for Life (Guido van Rossum) to a steering council illustrates how mature projects institutionalize decision-making.

These governance frameworks must balance inclusive participation with technical coherence, preventing fragmentation while encouraging innovation. Recent controversies around project leadership transitions, code of conduct enforcement, and corporate influence highlight the ongoing evolution of open-source governance as projects scale in importance and contributor diversity. The emergence of OpenSSF (Open Source Security Foundation) demonstrates how the ecosystem develops new structures to address systemic challenges like supply chain security.

5.4.3 Security Challenges in the Open-Source Supply Chain

The widespread adoption of open-source components has exposed critical vulnerabilities in the software supply chain, with high-profile incidents like Log4j demonstrating systemic risks. Many essential open-source projects rely on under-resourced maintainers, creating security blind spots in widely used dependencies. The 2021 White House Executive Order on Cybersecurity specifically addressed open-source security, leading to initiatives like Google's $100 million investment in open-source maintenance.

Package managers like npm, PyPI, and Maven Central have become single points of failure, vulnerable to typosquatting and dependency confusion attacks. Solutions emerging include Sigstore for cryptographic signing, GUAC for dependency graph analysis, and Alpha-Omega's critical project support program. These efforts reflect a growing recognition that open-source security is a global public good requiring sustained investment and coordination between maintainers, corporations, and governments.

5.4.4 Corporate Engagement and Open-Source Sustainability

The relationship between commercial entities and open-source communities has evolved into a complex symbiosis with inherent tensions. Tech giants now employ thousands of developers to work on open-source projects while facing criticism for extracting disproportionate value. The "tragedy of the commons" looms as companies benefit from open-source infrastructure without adequately supporting maintenance.

Models like Open Core (offering proprietary extensions) and SaaS implementations attempt to reconcile sustainability with open-source ideals. GitHub Sponsors, Open Collective, and Tidelift represent financial mechanisms to fund maintainers, though most critical projects remain underfunded. The rise of open-source program offices (OSPOs) in enterprises indicates growing recognition that strategic open-source engagement requires dedicated management and policy frameworks. This evolving dynamic continues to reshape how value flows through the open-source ecosystem.

5.4.5 Legal and Licensing Landscape Evolution

Open-source licensing has grown increasingly nuanced as the ecosystem matures and confronts new technological realities. The GPL's strong copyleft provisions dominated early open-source but face challenges in cloud-native environments where SaaS providers avoid distribution requirements. Permissive licenses (MIT, Apache 2.0) now dominate new projects, reflecting developer preferences for maximum adoption. Emerging issues include license compatibility with AI training data, patent clauses in corporate-contributed projects, and enforcement in international jurisdictions. The Software Package Data Exchange (SPDX) standard helps organizations manage license compliance at scale, while initiatives like OpenChain establish best practices for open-source license management. Recent court rulings on API copyrightability and the European Union's Cyber Resilience Act demonstrate how legal systems increasingly recognize open-source's unique characteristics while imposing new obligations.

5.4.6 Global Participation and Digital Divide Challenges

While open-source development theoretically offers equal participation opportunities, significant disparities persist in global contribution patterns. North America, Europe, and parts of Asia dominate project leadership, with underrepresented regions often relegated to consumer rather than creator roles. Barriers include language dominance (English as lingua franca), time zone challenges for collaboration, and unequal access to high-speed internet. Organizations like Outreachy and Google Summer of Code attempt to broaden participation, while GitHub's Copilot raises questions about who benefits from aggregated open-source knowledge. The growing practice of "open-source forking as protest" demonstrates how open-source intersects with international relations. Truly globalizing the open-source ecosystem requires addressing these structural inequities in access, recognition, and decision-making power.

5.4.7 Education and Workforce Development Impacts

Open-source methodologies have transformed technology education and professional development worldwide. Computer science programs increasingly incorporate open-source contribution as pedagogy, teaching collaboration alongside coding skills. Platforms like freeCodeCamp and The Odin Project provide open-source-aligned curricula accessible globally. The Linux Foundation's certification programs and Red Hat's training initiatives demonstrate how open-source skills have become career differentiators. However, credentialing challenges persist as employers struggle to evaluate unconventional contribution histories. The rise of "hacktoberfest" events and corporate-sponsored contribution programs indicates growing recognition that open-source experience develops valuable workplace competencies beyond technical ability—including remote collaboration, documentation, and community management.

5.5 TIMELINES AND MIGRATION PLANNING

The transition to post-quantum cryptography represents one of the most significant security challenges of the digital age, requiring careful planning and coordinated execution across industries and governments. As quantum computing advances toward practical realization, organizations must prepare to migrate their cryptographic infrastructure from classical algorithms to quantum-resistant alternatives. This migration process involves not just technical implementation but also strategic planning to ensure continuity of operations while maintaining security throughout the transition. The complexity of this undertaking stems from the pervasive nature of cryptographic systems in modern infrastructure, where encryption protects everything from financial transactions to national security communications. Effective migration planning requires understanding both the technical aspects of cryptographic algorithms and the operational realities of large-scale system updates.

The urgency of this migration becomes apparent when considering the potential consequences of delayed action. Quantum computers capable of breaking classical cryptography could compromise encrypted communications, digital signatures, and authentication mechanisms that secure modern digital infrastructure. The "harvest now, decrypt later" threat scenario, where adversaries collect encrypted data today to decrypt once quantum computers become available, underscores the need for proactive migration planning. Organizations must develop comprehensive timelines that account for the various phases of cryptographic transition, from initial assessment to full deployment, while considering the interdependencies between different systems and components.

The migration to post-quantum cryptography represents a complex but essential undertaking for maintaining secure digital infrastructure in the quantum computing era. Effective planning requires comprehensive assessments of cryptographic inventories, development of phased implementation strategies, and careful consideration of interoperability requirements. The timeline for this transition must balance the urgency of quantum threats with the operational realities of large-scale system updates, incorporating testing phases and resource allocations that support successful migrations.

The lessons learned from this transition will extend beyond the immediate migration process to inform future cryptographic updates. As organizations navigate this challenging but necessary evolution of their security infrastructure, the experiences gained will shape approaches to subsequent cryptographic transitions. The migration to post-quantum cryptography ultimately serves as both a response to emerging threats and an investment in the long-term security of digital communications, ensuring that organizations can maintain confidentiality and integrity in an era of advancing computational capabilities.

5.5.1 Quantum Threat Timeline

Estimating the timeline for practical quantum computing capabilities presents significant challenges, as progress depends on both scientific breakthroughs and engineering advancements. Current projections suggest that cryptographically relevant quantum computers capable of breaking widely used algorithms like

RSA and ECC could emerge within the next decade, though significant uncertainty remains. The development timeline may accelerate as research advances or encounter obstacles that delay practical realization. This uncertainty necessitates a proactive approach to cryptographic migration, where organizations prepare for quantum threats before they materialize rather than reacting after vulnerabilities become exploitable.

The timeline for quantum computing development also varies across different threat scenarios. While breaking public-key cryptography represents the most immediate concern, symmetric algorithms face different quantum threats that may require adjustments on a different schedule. Organizations must consider these varying timelines when planning their migration strategies, prioritizing the most vulnerable components while developing comprehensive plans for complete cryptographic transitions.

The potential for incremental quantum computing advances that could weaken cryptographic security before full-scale attacks become possible further complicates timeline planning, requiring organizations to remain vigilant throughout the migration process.

5.5.2 Assessing Cryptographic Inventories

The foundation of effective migration planning lies in comprehensive cryptographic inventories that identify all systems, protocols, and applications relying on classical algorithms. This assessment process involves cataloging cryptographic assets across an organization's entire digital infrastructure, including both obvious components like TLS implementations and less apparent elements such as legacy systems or third-party integrations.

The complexity of modern IT environments, with their interconnected systems and layered dependencies, makes this inventory process particularly challenging. Organizations must develop methodologies for identifying cryptographic usage that extend beyond surface-level examinations to uncover embedded or indirect dependencies. The inventory process should classify cryptographic assets based on their criticality and the urgency of migration.

Systems protecting highly sensitive data or supporting critical infrastructure require prioritization in migration planning. The assessment should also consider the lifecycle of different systems, as components nearing end-of-life may require different migration approaches than newer implementations. This classification process helps organizations develop phased migration plans that address the most vulnerable elements first while ensuring comprehensive coverage across all cryptographic assets. The inventory results form the basis for subsequent migration planning, guiding resource allocation and timeline development.

5.5.3 Developing Migration Strategies

Effective migration strategies balance the need for comprehensive cryptographic updates with the operational realities of maintaining business continuity. Organizations must develop approaches that account for the interdependencies between different systems, where updates to one component may require corresponding changes in others. The strategy development process involves evaluating different migration pathways, including hybrid approaches that combine classical and post-quantum algorithms during transition periods. These strategies should incorporate contingency plans that address potential implementation challenges or unforeseen compatibility issues that may arise during the migration process.

The migration strategy should also consider the human and organizational aspects of cryptographic transitions. Training programs for IT staff and end-users help ensure smooth adoption of new cryptographic systems. Change management processes that account for organizational culture and operational workflows can facilitate more effective migrations. The strategy should include mechanisms for monitoring progress and identifying potential issues throughout the transition, allowing for adjustments as needed. By developing comprehensive migration strategies that address both technical and organizational considerations, enterprises can navigate the complex process of cryptographic updates while minimizing disruptions to operations.

5.5.4 Phased Implementation Approaches

The complexity of cryptographic migration necessitates phased implementation approaches that break the transition into manageable components. Initial phases typically focus on the most vulnerable systems and those protecting critical assets, where the consequences of quantum attacks would be most severe. These early phases may involve pilot implementations that test post-quantum algorithms in controlled environments before broader deployment. Subsequent phases can address less critical systems while incorporating lessons learned from initial implementations. This incremental approach allows organizations to validate new cryptographic systems and refine migration processes before full-scale deployment.

The phased approach should incorporate specific milestones that mark progress through the migration timeline. These milestones help organizations track implementation status and ensure that the transition remains on schedule. Each phase should include validation processes that verify the security and functionality of new cryptographic implementations before proceeding to subsequent stages. The phased approach also allows for resource allocation that matches organizational capacities, preventing overburdening of IT departments during the transition. By structuring the migration as a series of manageable phases, organizations can maintain operational continuity while systematically addressing cryptographic vulnerabilities.

5.5.5 Interoperability and Compatibility Considerations

Migration planning must account for the interoperability requirements that arise from cryptographic transitions in interconnected systems. Organizations frequently operate in environments where multiple entities must communicate securely, requiring coordinated cryptographic updates across different platforms and jurisdictions. The migration strategy should identify potential compatibility issues that may emerge when some systems transition to post-quantum algorithms while others remain on classical cryptography.

These considerations become particularly important in supply chain relationships, where different organizations may progress through migration timelines at varying paces. Addressing interoperability challenges often involves developing transition mechanisms that support multiple cryptographic algorithms simultaneously. Hybrid systems that can process both classical and post-quantum cryptography provide bridges during migration periods. The planning process should also consider the development of cryptographic agility capabilities that allow systems to switch between different algorithms as needed. These approaches help maintain secure communications throughout the transition while accommodating varying migration timelines across interconnected systems. By proactively addressing interoperability considerations, organizations can prevent disruptions to collaborative workflows during cryptographic updates.

5.5.6 Testing and Validation Processes

Comprehensive testing and validation form critical components of cryptographic migration planning, ensuring that new implementations meet security requirements while maintaining operational functionality. The testing process should begin with laboratory evaluations that assess post-quantum algorithms in controlled environments before progressing to pilot implementations in production-like settings. These tests should evaluate not just the cryptographic security of new algorithms but also their performance characteristics and compatibility with existing systems. The validation process should incorporate both automated testing procedures and manual reviews to identify potential issues that might not be apparent through standard evaluation methods.

The testing timeline should allow for iterative refinement of cryptographic implementations based on evaluation results. Organizations should develop rollback procedures that can revert to classical algorithms if significant issues emerge during testing phases. The validation process should also include security assessments that consider potential new attack vectors introduced by post-quantum algorithms or their implementations. By allocating sufficient time for thorough testing and validation, organizations can identify and address implementation challenges before full deployment, reducing the risk of operational disruptions or security vulnerabilities during the migration process.

5.5.7 Resource Allocation and Timeline Management

Effective migration planning requires careful resource allocation that accounts for the personnel, technological, and financial requirements of cryptographic transitions. Organizations must develop realistic timelines that balance the urgency of quantum threats with operational capacities for implementing changes.

The resource planning process should identify necessary skills and expertise, potentially requiring training programs or external consulting support to address gaps in internal capabilities. Budgetary considerations must account for both direct implementation costs and potential indirect expenses that may arise from system updates or compatibility adjustments.

The timeline management process should incorporate buffer periods that accommodate potential delays in the migration process. These buffers help prevent cascading schedule disruptions that could occur if initial phases encounter unforeseen challenges. The resource allocation should also consider the need for parallel operations during transition periods, where both classical and post-quantum systems may need to run concurrently. By developing comprehensive resource plans that account for the full scope of migration requirements, organizations can execute cryptographic transitions more effectively while minimizing strain on operational capacities.

CHAPTER 6

PQC Implementation and Engineering

6.1 SOFTWARE LIBRARIES (E.G., LIBOQS, PQCLEAN, BOUNCY CASTLE PQC)

6.1.1 Liboqs

LibOQS (Open Quantum Safe) is an open-source software library designed to facilitate the integration of post-quantum cryptography (PQC) into various applications. As quantum computing advances, traditional cryptographic algorithms such as RSA and ECC (Elliptic Curve Cryptography) face significant threats due to their vulnerability to quantum attacks. LibOQS addresses this challenge by providing a collection of quantum-resistant cryptographic algorithms, enabling developers to experiment with and adopt PQC in their systems. The library serves as a foundational tool for research and practical implementation, ensuring a smooth transition to quantum-secure cryptography.

LibOQS represents a critical step toward securing digital communications in the quantum era. By providing a robust and accessible implementation of post-quantum cryptographic algorithms, the library empowers developers to prepare for the threats posed by quantum computing. As standardization efforts progress and performance optimizations continue, LibOQS will play an increasingly vital role in ensuring the long-term security of cryptographic systems. Organizations and developers are encouraged to explore and adopt LibOQS to stay ahead in the evolving landscape of cybersecurity. The transition to post-quantum cryptography is not just a necessity but an opportunity to build a more secure and resilient digital future.

Despite its advantages, the adoption of post-quantum cryptography presents several challenges. The larger key sizes and higher computational demands of quantum-resistant algorithms may pose difficulties for resource-constrained devices. Moreover, transitioning from classical to post-quantum cryptography requires careful planning to maintain backward compatibility and interoperability. LibOQS aims to address these challenges by providing optimized implementations and facilitating gradual migration strategies. Future developments in the library will focus on improving performance, expanding algorithm support, and enhancing ease of integration.

Need for Post-Quantum Cryptography

The rise of quantum computing poses a substantial risk to classical cryptographic systems. Quantum computers, leveraging Shor's algorithm, can efficiently break widely used public-key cryptosystems, including RSA, Diffie-Hellman, and ECC. This capability threatens the security of encrypted communications, digital signatures, and authentication mechanisms. To mitigate these risks, researchers have been developing post-quantum cryptographic algorithms that remain secure even against quantum attacks. LibOQS plays a crucial role in this transition by offering a standardized and accessible implementation of these algorithms, allowing developers to test and integrate them into real-world applications.

Features and Components of LibOQS

LibOQS provides a comprehensive suite of post-quantum cryptographic primitives, including key encapsulation mechanisms (KEMs) and digital signature schemes. The library supports multiple PQC algorithms that have been submitted to the National Institute of Standards and Technology (NIST) Post-Quantum Cryptography Standardization Project. Some of the supported algorithms include Kyber, Dilithium, Falcon, and SPHINCS+, each offering different security and performance trade-offs. LibOQS is designed to be modular, allowing developers to select the most suitable algorithms based on their specific requirements. Additionally, the library includes wrappers for integration with popular cryptographic protocols such as TLS (Transport Layer Security) through projects like OpenSSL and WolfSSL.

Integration and Compatibility

One of the key strengths of LibOQS is its compatibility with existing cryptographic frameworks. The library provides APIs that enable seamless integration with applications requiring secure communication. For instance, LibOQS can be used alongside OpenSSL to enable quantum-resistant TLS handshakes, ensuring that data transmitted over networks remains protected against future quantum threats. Furthermore, LibOQS supports cross-platform development, making it accessible for use in various operating systems and environments. This flexibility ensures that developers can incorporate post-quantum cryptography into diverse applications, from web servers to IoT devices.

Performance Considerations

While post-quantum cryptographic algorithms offer enhanced security, they often come with increased computational overhead compared to classical algorithms. LibOQS addresses this challenge by optimizing implementations for performance without compromising security. The library includes benchmarks that allow developers to evaluate the efficiency of different algorithms in terms of key generation, encryption, and decryption speeds. These benchmarks help in selecting the most appropriate algorithm based on the performance constraints of a given application. Additionally, ongoing research and updates to LibOQS aim to further improve the efficiency of quantum-resistant cryptography, ensuring that it becomes practical for widespread adoption.

Security and Standardization Efforts

Security is a paramount concern in cryptographic implementations, and LibOQS adheres to rigorous standards to ensure robustness. The algorithms included in the library are based on well-studied mathematical problems that are believed to be resistant to quantum attacks.

Furthermore, LibOQS aligns with NIST's post-quantum cryptography standardization process, which involves extensive evaluation of candidate algorithms for security, efficiency, and practicality. By leveraging NIST's recommendations, LibOQS ensures that developers have access to vetted and reliable cryptographic solutions. The open-source nature of the library also allows for community scrutiny, further enhancing its security through peer review.

Use Cases and Applications

LibOQS is applicable in a wide range of scenarios where secure communication and data protection are critical. In the financial sector, the library can be used to secure transactions and digital signatures against future quantum threats. Government and defense organizations can employ LibOQS to protect classified communications and infrastructure. Additionally, cloud service providers and IoT manufacturers can integrate the library to safeguard data transmitted between devices and servers. The versatility of LibOQS makes it a valuable tool for any industry seeking to future-proof its cryptographic systems.

6.1.2 PQClean

PQClean stands for "Post-Quantum Clean," emphasizing the importance of clean and readable code. It is not simply a cryptographic toolkit; it is a collaborative effort maintained by contributors from around the world, including academic researchers and professional developers. The primary motivation behind PQClean is to produce a reference library where each algorithm's implementation adheres to clear guidelines for code structure, formatting, and portability.

Clean code practices ensure that the implementations are auditable, understandable, and maintainable, which is essential for security-critical software. Unlike other cryptographic libraries that may include assembly or platform-specific optimizations, PQClean focuses on pure C implementations with optional Rust wrappers. This approach allows for greater compatibility across platforms and simplifies integration with various cryptographic systems.

PQClean is a vital open-source library dedicated to the clean, secure, and portable implementation of post-quantum cryptographic algorithms. As an essential component of the NIST PQC standardization process and broader cryptographic research, PQClean provides trustworthy code that can be used in testing, deployment, and education. Its structured approach, community-driven model, and strong focus on correctness make it a model for how cryptographic software should be developed in the post-quantum era. As threats from quantum computing become more imminent, tools like PQClean will be indispensable in securing our digital future.

As the world moves toward the adoption of quantum-resistant cryptography, the role of PQClean will continue to grow. With the final round of the NIST PQC project nearing completion and the transition to standardized algorithms already underway, software libraries like PQClean will be essential in supporting secure implementation and deployment.

Future developments may include enhanced side-channel protections, broader language bindings (such as Rust and Python), and support for new hardware platforms. The maintainers of PQClean are expected to adapt and evolve the library in line with emerging industry standards and user needs. By maintaining its commitment to clarity, security, and openness, PQClean will remain a cornerstone in the global effort to build quantum-secure systems.

Role in NIST Post-Quantum Standardization

PQClean plays a vital supporting role in the NIST Post-Quantum Cryptography Standardization project. The project aims to evaluate and standardize quantum-resistant public-key cryptographic algorithms. Many of the candidate algorithms submitted to NIST have reference implementations included in PQClean. These implementations follow strict coding standards and are designed for cross-platform compatibility. This makes PQClean a useful tool for researchers, implementers, and evaluators who want to test the performance, correctness, and interoperability of the proposed algorithms. By serving as a bridge between academic innovation and practical deployment, PQClean contributes significantly to the broader goal of preparing cryptographic systems for a post-quantum future.

Design Philosophy and Code Structure

The design of PQClean prioritizes simplicity, clarity, and security. The codebase is organized in a modular format, where each cryptographic algorithm is contained within its own directory structure. Each algorithm directory contains subfolders for specific parameter sets and functional categories such as key generation, encryption, and decryption. Furthermore, each function is implemented in a consistent and reusable way across the library. This structured layout not only enhances readability but also facilitates automated testing and verification. Additionally, the naming conventions and file structures follow predictable patterns, allowing developers to easily navigate through the library and integrate only the components they need.

Support for Multiple Algorithm Families

PQClean supports a diverse set of post-quantum algorithm families. These include lattice-based schemes like Kyber and Dilithium, code-based schemes such as Classic McEliece, multivariate polynomial schemes like Rainbow, and hash-based signature schemes like SPHINCS+. Each of these families represents a different approach to achieving quantum resistance.

For example, Kyber is a key encapsulation mechanism based on the module learning with errors (MLWE) problem, while Dilithium is a digital signature scheme also based on lattice problems. Classic McEliece, on the other hand, relies on the hardness of decoding random linear codes. PQClean provides clean implementations of these schemes as submitted to NIST, helping the cryptographic community compare their security, performance, and implementation complexity in a unified environment.

Portability and Integration

One of PQClean's major strengths is its emphasis on portability. Because the code is written in standard C with minimal dependencies, it can be compiled and executed on a wide range of platforms, including Linux, Windows, macOS, and embedded systems. This ensures that developers working in different environments can use and test the algorithms without needing to reimplement or heavily modify the code. PQClean also integrates well with other cryptographic libraries. For instance, the implementations in PQClean serve as a backend for the PQCrypto module in liboqs (Open Quantum Safe), which further extends the usage of PQClean algorithms in secure protocols such as TLS and SSH. This interoperability makes PQClean a foundational layer in the post-quantum software stack.

Verification and Testing

Security is at the heart of PQClean, and extensive verification is built into its development process. The library includes a suite of test vectors and automated scripts that verify the correctness and consistency of each algorithm's implementation. Developers and users can run these tests to confirm that the algorithms produce the expected outputs under various conditions. Moreover, PQClean supports constant-time implementation practices to protect against side-channel attacks, which are a significant concern in cryptographic software. Although not all algorithms are inherently side-channel resistant, PQClean aims to document and promote the use of secure coding techniques wherever possible. Through such transparency and validation, PQClean builds trust in the algorithms it hosts.

Contribution and Community Involvement

PQClean is a community-driven project, hosted on GitHub, and welcomes contributions from developers, researchers, and students alike. The open development model ensures that improvements, bug fixes, and new algorithm implementations can be proposed, reviewed, and integrated in a collaborative and transparent manner. The maintainers enforce rigorous code review and follow best practices in version control, documentation, and issue tracking. This vibrant community engagement not only accelerates the development of post-quantum tools but also serves as an educational platform for those interested in learning how secure and efficient cryptographic software is developed.

Educational and Research Applications

In addition to its practical uses, PQClean is also a valuable resource in educational settings. Its clean codebase and documentation make it an excellent tool for teaching post-quantum cryptography concepts at the undergraduate and graduate levels. Students and researchers can use PQClean to study the algorithmic foundations of post-quantum schemes, analyze implementation trade-offs, and experiment with integration into real-world applications. For academic institutions and research labs, PQClean provides a ready-made environment to prototype new algorithms and evaluate their suitability for standardization and deployment.

6.1.3 Bouncy Castle PQC

The Bouncy Castle PQC library represents a significant contribution to the field of post-quantum cryptography, providing developers with the tools needed to begin transitioning to quantum-resistant security. As the cryptographic community prepares for the challenges posed by quantum computing, resources like Bouncy Castle play a crucial role in facilitating this transition.

The library's comprehensive implementations, secure design, and educational value make it an essential component of the cryptographic ecosystem as it evolves to meet the demands of the quantum era. The ongoing development of the Bouncy Castle PQC library reflects the dynamic nature of post-quantum cryptography, where new algorithms and optimizations continue to emerge.

As standards evolve and new cryptanalytic insights develop, the library will remain an important resource for implementing and understanding quantum-resistant security. The experiences gained through working with the library will help shape the future of cryptographic implementations, ensuring that security systems can adapt to meet the challenges of advancing computational capabilities.

In this transitional period, Bouncy Castle PQC serves as both a practical tool and an educational platform, helping prepare the cryptographic community for the quantum future.

Architecture of Bouncy Castle PQC

The Bouncy Castle PQC library builds upon the established architecture of the core Bouncy Castle framework while extending it to support quantum-resistant algorithms. This architecture maintains the library's characteristic modular design, allowing developers to integrate post-quantum cryptographic functions alongside classical algorithms within the same codebase. The implementation follows object-oriented principles that facilitate clean integration with existing systems, making it easier for developers to incorporate quantum-resistant security into their applications. The library's design emphasizes both security and usability, providing well-documented APIs that abstract much of the complexity inherent in post-quantum cryptographic operations.

A key aspect of the library's architecture is its support for multiple post-quantum algorithm families, reflecting the diversity of approaches being considered for standardization. This includes lattice-based cryptography, hash-based signatures, code-based encryption, and multivariate schemes, among others. The architecture allows for algorithm agility, where applications can potentially switch between different post-quantum primitives as standards evolve or new vulnerabilities are discovered. This flexibility is particularly valuable in the current transitional period of cryptographic development, where multiple algorithm candidates are still under evaluation and the security landscape continues to shift.

Supported Post-Quantum Algorithms

The Bouncy Castle PQC library provides implementations of several post-quantum algorithm candidates that have been proposed for standardization. Among these are lattice-based cryptographic schemes, which have gained significant attention for their strong security guarantees and relatively good performance characteristics. The library includes implementations of Kyber, a key encapsulation mechanism based on learning with errors (LWE) problems, and Dilithium, a digital signature scheme also derived from lattice-based constructions. These algorithms have been selected as finalists in the NIST post-quantum standardization process, making their inclusion in Bouncy Castle particularly valuable for developers preparing for the quantum transition.

In addition to lattice-based schemes, the library supports other post-quantum algorithm families that demonstrate different approaches to quantum resistance. This includes hash-based signature schemes like SPHINCS+, which provides strong security guarantees based on well-understood cryptographic hash functions.

The library also implements code-based cryptographic constructions, offering developers exposure to the diversity of mathematical approaches being considered for post-quantum security. This comprehensive coverage of algorithm families allows developers to experiment with different quantum-resistant approaches and gain practical experience with their implementation characteristics.

Implementation Considerations and Security

The Bouncy Castle PQC library places significant emphasis on secure implementation practices, recognizing that the security of cryptographic algorithms depends not just on their mathematical foundations but also on their practical realization. The library's implementations follow best practices for constant-time execution and side-channel resistance, addressing common vulnerabilities that can compromise cryptographic security. This attention to implementation security is particularly important for post-quantum algorithms, many of which are still relatively new and may have implementation characteristics that differ from classical cryptographic primitives.

The library's development process incorporates extensive testing and validation to ensure that implementations meet security requirements. This includes both functional testing to verify correct operation and security testing to identify potential vulnerabilities. The open-source nature of the library allows for community scrutiny of implementations, where security researchers can examine the code for potential issues. This transparency helps build confidence in the security of the implementations while also providing opportunities for continuous improvement through community contributions. The combination of careful implementation practices and community review makes Bouncy Castle PQC a valuable resource for secure post-quantum cryptographic development.

Integration with Existing Systems

One of the most valuable aspects of the Bouncy Castle PQC library is its design for integration with existing cryptographic systems. The library's architecture allows developers to incorporate post-quantum algorithms alongside classical cryptographic functions, facilitating hybrid implementations that can support both algorithm families. This capability is particularly important during the transition period to post-quantum cryptography, where systems may need to maintain compatibility with classical algorithms while gradually adopting quantum-resistant primitives.

The library provides interfaces that abstract much of the complexity of post-quantum cryptographic operations, making it easier for developers to integrate these new algorithms into their applications. This includes support for common cryptographic operations like key generation, encryption, decryption, and digital signatures, implemented through familiar API patterns that mirror classical cryptographic functions. The design also accommodates the larger key sizes and different parameter requirements that characterize many post-quantum algorithms, handling these aspects transparently for the developer. These integration features make Bouncy Castle PQC a practical tool for developers beginning to work with quantum-resistant cryptography.

Performance Characteristics and Optimization

The Bouncy Castle PQC library addresses the performance considerations that are particularly relevant for post-quantum cryptographic algorithms. Many quantum-resistant primitives have different performance characteristics than classical algorithms, often requiring more computational resources or larger key sizes. The library's implementations include optimizations that improve the efficiency of these operations while maintaining security. This includes algorithm-specific optimizations that leverage the mathematical structure of particular post-quantum schemes to enhance performance.

The library also provides mechanisms for developers to make informed decisions about performance trade-offs. This includes documentation of the computational requirements and memory usage characteristics of different algorithms, allowing developers to select appropriate primitives based on their application requirements. For resource-constrained environments, the library offers guidance on parameter selection and implementation strategies that can help mitigate performance impacts. These considerations are particularly important as post-quantum algorithms are deployed in diverse environments, from high-performance servers to embedded systems.

Educational Value and Community Contributions

Beyond its practical implementation value, the Bouncy Castle PQC library serves an important educational role in the cryptographic community. The open-source nature of the library, combined with its comprehensive documentation, makes it an accessible resource for learning about post-quantum cryptographic algorithms. Developers can examine the implementations to understand how different quantum-resistant schemes operate, gaining insights into their mathematical foundations and practical characteristics. This educational aspect is particularly valuable given the complexity of many post-quantum algorithms and the relative newness of the field.

The library also benefits from community contributions that enhance its capabilities and security. The open-source development model allows for continuous improvement through community feedback, bug reports, and code contributions. This collaborative approach helps ensure that the library remains current with the latest developments in post-quantum cryptography, incorporating new algorithms and optimizations as they emerge. The community aspect also provides opportunities for peer review of implementations, where security researchers can examine the code for potential vulnerabilities, contributing to the overall security of the library.

Challenges and Considerations in PQC Implementation

The implementation of post-quantum cryptographic algorithms presents several challenges that the Bouncy Castle PQC library helps address. One significant challenge is the larger parameter sizes that characterize many quantum-resistant schemes, which can impact both performance and storage requirements. The library's design considers these aspects, providing guidance on parameter selection and implementation strategies that can help mitigate these impacts. Another challenge involves the different security assumptions and implementation characteristics of post-quantum algorithms compared to classical primitives, which the library helps developers navigate through its comprehensive documentation and API design.

The library also addresses the challenge of algorithm agility, where systems may need to support multiple cryptographic primitives or transition between them. The modular design of Bouncy Castle PQC facilitates this flexibility, allowing developers to incorporate different post-quantum algorithms and potentially switch between them as standards evolve or new cryptanalytic insights emerge. This capability is particularly important in the current transitional period of cryptographic development, where multiple algorithm candidates are still under evaluation and the security landscape continues to shift.

Role of Bouncy Castle in Cryptographic Transition

The Bouncy Castle PQC library plays a crucial role in the broader transition to post-quantum cryptography by providing developers with practical tools for implementing quantum-resistant security. As organizations prepare for the quantum computing era, the library serves as both an educational resource and a practical implementation framework. This dual role helps bridge the gap between theoretical cryptographic research and real-world deployment, allowing developers to gain experience with post-quantum algorithms before they become formal standards.

The library's comprehensive nature, covering multiple algorithm families and providing secure implementations, makes it an invaluable resource for the cryptographic community. By offering open-source implementations of post-quantum algorithms, Bouncy Castle enables broader experimentation and adoption of these new security mechanisms. This contributes to the overall maturation of post-quantum cryptography, helping identify implementation best practices and potential vulnerabilities through community use and scrutiny. The role of Bouncy Castle in this transition underscores the importance of open-source cryptographic libraries in advancing the state of security technology.

6.2 HARDWARE IMPLEMENTATIONS AND ACCELERATION (FPGAS, ASICS)

The hardware implementation and acceleration of post-quantum cryptographic algorithms represent essential components of the transition to quantum-resistant security. FPGAs and ASICs provide the performance improvements necessary for practical deployment of quantum-resistant algorithms across diverse technological environments. The careful optimization of hardware designs balances performance requirements with security considerations, producing implementations that can meet the demanding needs of modern cryptographic applications. As organizations prepare for the quantum computing era, hardware acceleration emerges as a crucial enabler for maintaining secure communications and data protection.

The development of hardware-accelerated post-quantum cryptography reflects the cryptographic community's recognition that software implementations alone may be insufficient for meeting the performance requirements of many real-world applications. The innovative approaches to hardware design and implementation demonstrate the field's adaptability in addressing emerging security challenges. As post-quantum cryptographic standards continue to evolve, hardware implementations will play an increasingly important role in enabling secure and efficient cryptographic operations. The experiences gained through current hardware acceleration efforts will inform future developments, ensuring that cryptographic systems can adapt to meet the challenges of advancing computational capabilities.

The future of hardware-accelerated post-quantum cryptography involves several promising directions that will shape the development of quantum-resistant security solutions. The integration of hardware acceleration with emerging computational paradigms such as quantum computing presents new opportunities for cryptographic implementations. Hybrid systems that combine classical and quantum computational elements may offer novel approaches to post-quantum security. The development of specialized hardware architectures optimized for specific post-quantum algorithm families will likely continue, with designs becoming increasingly tailored to particular mathematical structures.

Advancements in hardware security features will play a crucial role in future cryptographic implementations. New techniques for side-channel resistance and fault injection protection will emerge as physical attacks become more sophisticated. The integration of hardware acceleration with secure enclave technologies may provide comprehensive protection for cryptographic operations. As post-quantum cryptographic standards mature, hardware implementations will likely become increasingly specialized, with designs optimized for specific algorithms and deployment scenarios. These future directions will shape the evolution of hardware-accelerated cryptography, ensuring that security systems can meet the challenges of advancing computational capabilities.

6.2.1 Fundamentals of Hardware Acceleration for Cryptography

Hardware acceleration of cryptographic operations leverages the parallel processing capabilities and specialized architectures of hardware platforms to improve computational efficiency. FPGAs provide reconfigurable hardware that can be optimized for specific cryptographic algorithms, offering flexibility in implementation while delivering substantial performance improvements. ASICs, on the other hand, offer dedicated hardware implementations that can achieve even greater efficiency but with less flexibility for modification. Both approaches exploit the inherent parallelism in many post-quantum algorithms, where operations can be executed concurrently to reduce overall computation time.

The design of hardware-accelerated cryptographic implementations involves careful consideration of algorithm characteristics and hardware capabilities. Many post-quantum algorithms exhibit mathematical structures that lend themselves particularly well to hardware acceleration. For instance, lattice-based cryptographic operations often involve matrix multiplications and polynomial arithmetic that can be efficiently parallelized in hardware. The hardware implementation process requires mapping these algorithmic structures to hardware architectures in ways that maximize performance while maintaining security. This involves not just the cryptographic operations themselves but also the memory access patterns and data flow characteristics that can significantly impact overall performance.

6.2.2 FPGA Implementations of Post-Quantum Cryptography

Field-Programmable Gate Arrays have emerged as particularly valuable platforms for accelerating post-quantum cryptographic operations due to their combination of performance and flexibility. FPGA implementations allow for the optimization of cryptographic algorithms through custom hardware designs that can be reconfigured as needed. This reconfigurability is especially valuable in the current transitional period of post-quantum cryptography, where multiple algorithm candidates are still under evaluation and standards continue to evolve. Developers can update FPGA implementations to support new algorithms or modified parameters as cryptographic requirements change.

The implementation of post-quantum algorithms on FPGAs typically involves several optimization techniques to maximize performance. Pipelining allows for the concurrent execution of multiple operations, significantly improving throughput for algorithms with sequential dependencies. Parallel processing exploits the inherent parallelism in many post-quantum operations, executing independent computations simultaneously. Memory optimization techniques address the often substantial memory requirements of quantum-resistant algorithms, implementing efficient storage and access patterns. These optimization strategies, combined with the reconfigurable nature of FPGAs, make them particularly suitable for accelerating post-quantum cryptographic operations in diverse deployment scenarios.

6.2.3 ASIC Implementations and Their Advantages

Application-Specific Integrated Circuits offer even greater performance improvements for post-quantum cryptographic operations through dedicated hardware implementations. ASIC designs can be precisely optimized for specific cryptographic algorithms, achieving efficiency levels that surpass what is possible with FPGAs or software implementations. This optimization potential makes ASICs particularly attractive for high-volume applications where performance is critical, such as in data centers or specialized security hardware. The fixed nature of ASIC implementations, however, requires careful consideration of algorithm selection and parameter choices to ensure long-term viability.

The development of ASIC implementations for post-quantum cryptography involves specialized design techniques that exploit the mathematical characteristics of quantum-resistant algorithms. Custom arithmetic units can be designed to efficiently execute the specific operations required by particular algorithms, such as polynomial multiplication in lattice-based schemes. Memory architectures can be optimized for the access patterns of post-quantum operations, reducing latency and improving throughput. The physical design of ASICs can also incorporate security features that protect against side-channel attacks, which are particularly relevant for cryptographic implementations. These dedicated hardware solutions offer substantial performance advantages but require significant development efforts and careful planning to ensure their continued relevance as cryptographic standards evolve.

6.2.4 Performance Considerations

The performance characteristics of hardware-accelerated post-quantum cryptographic implementations depend on several interrelated factors that must be carefully balanced. Throughput and latency represent primary performance metrics that often involve trade-offs in hardware design. High-throughput implementations that can process multiple cryptographic operations concurrently may introduce additional latency for individual operations. Conversely, low-latency designs optimized for minimal processing delay may limit overall throughput. The specific performance requirements of different applications influence these design decisions, with some scenarios prioritizing high throughput while others demand minimal latency.

Resource utilization forms another critical performance consideration in hardware implementations. The computational and memory resources required by post-quantum algorithms often exceed those of classical cryptographic primitives, necessitating careful optimization in hardware designs. Power consumption becomes particularly important for embedded and mobile applications, where energy efficiency can be as critical as computational performance. The physical characteristics of hardware implementations, including chip area and thermal properties, also influence performance considerations. These interrelated factors require comprehensive evaluation in the design of hardware-accelerated post-quantum cryptographic systems to ensure they meet the specific requirements of their intended deployment environments.

6.2.5 Security Considerations

The security of hardware-accelerated cryptographic implementations extends beyond the mathematical security of the underlying algorithms to include protection against physical attacks. Side-channel resistance represents a particularly important consideration for hardware implementations, where physical characteristics of the hardware can potentially leak sensitive information. Power analysis attacks that monitor energy consumption patterns and electromagnetic attacks that measure radiation emissions can compromise cryptographic security if not properly addressed. Hardware implementations must incorporate countermeasures such as constant-time execution and masking techniques to protect against these physical attack vectors.

Fault injection attacks present another security consideration for hardware cryptographic implementations. These attacks involve intentionally inducing errors in hardware operations to extract secret information or bypass security mechanisms. Robust error detection and correction mechanisms can help mitigate these threats. The physical security of hardware implementations also requires consideration of tamper resistance and secure storage mechanisms for cryptographic keys and sensitive data. These security considerations must be addressed throughout the hardware design process, from architectural decisions to physical layout, to ensure that performance improvements do not come at the expense of cryptographic security.

6.2.6 Design Methodologies

The development of hardware-accelerated post-quantum cryptographic implementations follows systematic design methodologies that balance performance, security, and practical considerations. High-level synthesis approaches allow for the exploration of architectural trade-offs before committing to specific hardware implementations. This involves modeling cryptographic algorithms at higher levels of abstraction to evaluate different design options. The co-design of hardware and software components enables optimization across the entire system, considering how hardware acceleration integrates with overall cryptographic implementations.

Verification methodologies form a crucial component of hardware design processes, ensuring that implementations meet both functional and security requirements. Formal verification techniques can mathematically prove the correctness of hardware designs, while simulation-based approaches evaluate performance characteristics and identify potential vulnerabilities. The iterative nature of hardware design allows for refinement based on verification results, improving both security and performance through successive development cycles. These comprehensive design methodologies help produce hardware implementations that meet the demanding requirements of post-quantum cryptographic applications.

6.2.7 Challenges

The implementation of post-quantum cryptographic algorithms in hardware presents several unique challenges that distinguish them from classical cryptographic primitives. The often larger parameter sizes of quantum-resistant algorithms can strain memory resources and bandwidth requirements, necessitating innovative architectural solutions. The mathematical complexity of many post-quantum operations can make efficient hardware implementation particularly challenging, requiring careful optimization of computational units. The relative newness of many post-quantum algorithms means that hardware designers often lack the extensive implementation experience available for classical cryptographic primitives.

The evolving nature of post-quantum cryptographic standards presents additional challenges for hardware implementation. As algorithm candidates progress through standardization processes, their specifications may change, potentially requiring hardware redesigns. The need to support multiple algorithm candidates during transitional periods can complicate hardware implementations that are typically optimized for specific operations. These challenges require flexible design approaches that can accommodate potential changes while still delivering substantial performance improvements. The hardware implementation community must navigate these challenges while developing solutions that can be practically deployed in real-world systems.

6.2.8 Case Studies

Several case studies demonstrate the practical implementation of hardware-accelerated post-quantum cryptographic algorithms across different platforms and applications. FPGA implementations of lattice-based cryptographic schemes have shown substantial performance improvements through parallel processing of polynomial arithmetic operations. These implementations leverage the reconfigurable nature of FPGAs to support multiple algorithm candidates, providing flexibility during the standardization transition period. The performance characteristics of these implementations have been evaluated across different parameter sets, demonstrating the trade-offs between security and computational efficiency.

ASIC implementations of hash-based signature schemes have achieved remarkable throughput improvements through dedicated hardware designs optimized for specific cryptographic operations. These implementations often incorporate custom memory architectures that efficiently handle the large data structures characteristic of many post-quantum algorithms.

The physical security features of these ASIC designs have been evaluated against various side-channel attack vectors, demonstrating the effectiveness of integrated countermeasures. These case studies provide valuable insights into the practical considerations of hardware-accelerated post-quantum cryptography, informing future implementation efforts.

6.3 PERFORMANCE CONSIDERATIONS: SPEED, SIZE, BANDWIDTH

The performance considerations of speed, size, and bandwidth in post-quantum cryptography represent critical factors influencing the adoption of quantum-resistant security solutions. As organizations prepare for the transition to post-quantum algorithms, comprehensive evaluation of these performance characteristics becomes essential for effective migration planning.

The complex trade-offs between security guarantees and operational efficiency require careful analysis and optimization strategies tailored to specific deployment environments. The experiences gained through current implementation efforts will inform future developments, ensuring that cryptographic systems can adapt to meet the challenges of advancing computational capabilities while maintaining robust security guarantees.

The future of post-quantum cryptography lies in the balanced development of security and performance characteristics that address diverse operational requirements. As quantum computing capabilities continue to advance, the cryptographic community's focus on performance optimization will remain crucial for enabling practical deployment of quantum-resistant security solutions.

The ongoing evolution of post-quantum algorithms and their implementations will shape the cryptographic landscape of the quantum era, where both security and performance considerations must be properly balanced to meet the demands of modern digital infrastructure.

The future of performance optimization in post-quantum cryptography involves several promising directions that will shape the evolution of quantum-resistant security solutions. Algorithm development that focuses on computational efficiency while maintaining security guarantees represents an important area of ongoing research. Hardware acceleration through specialized processors or co-processors may provide substantial performance improvements for specific cryptographic operations. Protocol design innovations that minimize communication overhead while preserving security properties could significantly enhance network performance.

The optimization directions extend to various aspects of cryptographic implementation. Adaptive algorithms that can adjust their performance characteristics based on available resources or specific operational requirements may offer improved efficiency. Hybrid implementations that combine different post-quantum approaches to optimize overall performance represent another promising direction. Machine learning techniques that identify optimization opportunities in cryptographic operations could provide new approaches to performance enhancement. These future directions will shape the development of post-quantum cryptographic solutions, ensuring that security systems can meet both current and emerging performance requirements in an evolving technological landscape.

6.4 SIDE-CHANNEL RESISTANCE AND IMPLEMENTATION SECURITY

The consideration of side-channel resistance and implementation security in post-quantum cryptography represents a crucial aspect of the transition to quantum-resistant security. As organizations prepare for the adoption of post-quantum algorithms, comprehensive attention to physical security considerations becomes essential for maintaining robust protection against both mathematical and implementation attacks. The complex nature of many quantum-resistant schemes, combined with their relative newness compared to classical cryptographic primitives, makes implementation security a particularly important aspect of their deployment. The future of post-quantum cryptography lies in the balanced development of algorithmic security and implementation protection that addresses diverse physical attack vectors. As computational capabilities continue to advance, the cryptographic community's focus on implementation security will remain crucial for enabling practical deployment of quantum-resistant security solutions. The experiences gained through current implementation efforts will inform future developments, ensuring that cryptographic systems can adapt to meet the challenges of advancing physical attack capabilities while maintaining robust security guarantees.

The future of implementation security in post-quantum cryptography involves several promising directions that will shape the development of quantum-resistant security solutions. The integration of hardware security features with cryptographic implementations presents opportunities for comprehensive protection against physical attacks. The development of specialized hardware architectures optimized for secure execution of post-quantum algorithms may provide enhanced protection while maintaining performance characteristics. The evolution of side-channel resistant programming techniques will continue to improve implementation security as new attack vectors emerge.

The advancement of formal verification techniques for cryptographic implementations may provide more rigorous guarantees of security characteristics. The development of automated testing methodologies that can identify potential side-channel vulnerabilities will enhance implementation security. The integration of machine learning techniques for analyzing physical execution characteristics may provide new approaches to detecting and mitigating implementation vulnerabilities. These future directions will shape the evolution of post-quantum cryptographic implementations, ensuring that security systems can meet both current and emerging physical attack challenges in an evolving technological landscape.

6.5 MIGRATION TOOLS AND APIS

The development and deployment of migration tools and APIs for post-quantum cryptography represent critical components of the transition to quantum-resistant security. These tools and interfaces provide the necessary capabilities to analyze existing cryptographic implementations, plan effective migration strategies, and execute updates to new algorithms while maintaining security and operational continuity. The comprehensive nature of these solutions addresses the complex challenges of cryptographic transitions, from inventory and analysis to deployment and validation.

The future of post-quantum cryptography lies in the balanced development of migration capabilities that address both technical and operational considerations. As organizations prepare for the quantum computing era, the experiences gained through current migration efforts will inform future developments, ensuring that cryptographic systems can adapt to meet the challenges of advancing computational capabilities. The ongoing evolution of migration tools and APIs will play a crucial role in enabling secure transitions to quantum-resistant cryptography, helping maintain robust protection in an era of rapidly advancing technological capabilities.

The future development of migration tools and APIs for post-quantum cryptography involves several promising directions that will shape the evolution of quantum-resistant security solutions. The integration of artificial intelligence techniques presents opportunities for more sophisticated cryptographic analysis and migration planning. Machine learning algorithms could help identify patterns in cryptographic usage that might indicate potential migration challenges or optimization opportunities. The development of automated migration capabilities that can update cryptographic implementations with minimal manual intervention represents another promising direction.

The evolution of cryptographic APIs will likely focus on enhanced abstraction capabilities that further simplify the adoption of post-quantum algorithms. The development of standardized interfaces that can support multiple cryptographic schemes through common APIs may facilitate broader adoption. The integration of performance monitoring and adaptive optimization features will help ensure that post-quantum implementations maintain operational efficiency. These future directions will shape the development of migration tools and APIs, ensuring that organizations can effectively transition to quantum-resistant cryptography while maintaining robust security guarantees and operational continuity.

The design of application programming interfaces for post-quantum cryptography follows specific principles that facilitate secure and efficient adoption of quantum-resistant algorithms. These APIs typically emphasize abstraction to hide the complexity of underlying cryptographic operations while providing standardized interfaces that developers can use consistently across different implementations. The abstraction layer allows applications to interact with cryptographic functions without needing to understand the specific details of post-quantum algorithms, making the transition process more manageable for development teams.

A key principle in API design involves maintaining compatibility with existing cryptographic interfaces to minimize the disruption caused by algorithm updates. This often involves creating wrapper functions that present familiar interfaces to applications while handling the underlying post-quantum operations. The APIs also typically incorporate algorithm agility features that allow for switching between different cryptographic primitives as standards evolve or new vulnerabilities are discovered. Security considerations form another fundamental principle, with APIs designed to prevent misuse that could compromise cryptographic protection. These design principles help ensure that post-quantum cryptographic APIs can be effectively integrated into existing systems while maintaining robust security guarantees.

CHAPTER 7

Transitioning to a Quantum-Safe World

7.1 CRYPTOGRAPHIC AGILITY: PRINCIPLES AND PRACTICES

The principles and practices of cryptographic agility represent essential components of the transition to quantum-resistant security. As organizations prepare for the quantum computing era, the ability to adapt cryptographic systems to evolving threats and standards becomes increasingly crucial. The comprehensive nature of cryptographic agility addresses the complex challenges of algorithm transitions, from maintaining security guarantees to preserving operational continuity. The experiences gained through current implementation efforts will inform future developments, ensuring that cryptographic systems can adapt to meet the challenges of advancing computational capabilities.

The future of cryptographic security lies in the balanced development of agility principles that address both technical and operational considerations. As quantum computing capabilities continue to advance, the cryptographic community's focus on adaptable systems will remain crucial for enabling practical deployment of quantum-resistant security solutions. The ongoing evolution of cryptographic agility will play a vital role in facilitating secure transitions to post-quantum cryptography, helping maintain robust protection in an era of rapidly advancing technological capabilities. The principles of cryptographic agility provide the necessary framework for navigating the complex transition to quantum-safe security while preserving the confidentiality and integrity of digital communications.

The future development of cryptographic agility involves several promising directions that will shape the evolution of adaptable security systems. The integration of artificial intelligence techniques presents opportunities for more sophisticated algorithm selection and parameter optimization. Machine learning approaches could help identify optimal cryptographic configurations based on specific security requirements and operational constraints. The development of automated migration capabilities that can update cryptographic implementations with minimal manual intervention represents another promising direction.

The evolution of cryptographic agility will likely focus on enhanced abstraction capabilities that further simplify the adoption of new algorithms. The development of standardized interfaces that can support an increasingly diverse set of cryptographic primitives may facilitate broader adoption. The integration of performance monitoring and adaptive optimization features will help ensure that cryptographic systems can maintain operational efficiency while accommodating evolving security requirements. These future directions will shape the development of cryptographic agility, ensuring that security systems can adapt to meet the challenges of advancing computational capabilities while maintaining robust protection in an evolving technological landscape.

7.1.1 Principles

The principles of cryptographic agility rest on several fundamental concepts that guide the development of adaptable security systems. Algorithm independence forms the cornerstone of these principles, where cryptographic implementations are designed to support multiple algorithms through standardized interfaces.

This approach enables the replacement of specific cryptographic primitives without requiring complete system redesigns. Parameter flexibility represents another key principle, allowing for adjustments to cryptographic parameters as security requirements evolve or new vulnerabilities are discovered. These principles help ensure that cryptographic systems can adapt to changing security landscapes while maintaining operational continuity.

The principles of cryptographic agility extend to various aspects of system design and implementation. Modular architecture forms an essential component, where cryptographic functions are implemented as interchangeable components that can be updated independently. Standardized interfaces provide consistent access to cryptographic operations regardless of the underlying algorithms, facilitating algorithm replacement. Configuration management capabilities allow for dynamic adjustment of cryptographic parameters based on evolving security requirements. These principles combine to create systems that can adapt to new cryptographic standards and threat landscapes while preserving security guarantees and operational functionality.

7.1.2 Role of Cryptographic Agility in Quantum Transition

Cryptographic agility plays a particularly crucial role in the transition to quantum-resistant security by providing the flexibility needed to navigate the current period of cryptographic evolution. As multiple post-quantum algorithm candidates remain under evaluation and new cryptanalytic insights continue to emerge, the ability to update cryptographic implementations becomes essential for maintaining security. The principles of cryptographic agility enable organizations to adopt new algorithms as standards evolve or vulnerabilities are discovered, ensuring that security systems remain protected against both classical and quantum threats. This capability helps mitigate the risks associated with cryptographic transitions by allowing for gradual adoption of new algorithms while maintaining compatibility with existing systems.

The transition to post-quantum cryptography involves not just the adoption of new algorithms but also the potential need to update cryptographic implementations as standards mature. Cryptographic agility provides the necessary framework for managing these updates without requiring complete system redesigns. The ability to support multiple algorithms simultaneously becomes particularly valuable during transition periods, where some systems may need to maintain compatibility with classical algorithms while gradually adopting quantum-resistant primitives. This flexibility helps ensure that organizations can navigate the complex process of cryptographic transitions while minimizing disruptions to operational continuity.

7.1.3 Architectural Considerations for Agile Systems

The development of cryptographic systems that embody the principles of agility requires careful consideration of architectural design. Modular cryptographic architectures form the foundation of these systems, where different cryptographic components can be updated independently. This modular approach involves designing cryptographic implementations as separate, interchangeable modules that communicate through standardized interfaces.

The architecture should support algorithm negotiation capabilities that allow communicating parties to agree on cryptographic algorithms based on their respective capabilities. This enables systems to maintain compatibility during transition periods when different components may support varying sets of cryptographic algorithms.

The architectural considerations extend to various aspects of system design and implementation. Configuration management systems that can dynamically adjust cryptographic parameters based on evolving security requirements form an essential component of agile architectures. The design should incorporate versioning capabilities that maintain backward compatibility with existing cryptographic implementations while supporting new algorithms. Error handling mechanisms that provide graceful degradation when compatibility issues arise help prevent disruptions during cryptographic transitions. These architectural considerations combine to create systems that can adapt to changing cryptographic standards while maintaining security guarantees and operational functionality.

7.1.4 Implementation Strategies

Effective implementation of cryptographic agility principles involves several strategies that facilitate the development of adaptable security systems. Algorithm abstraction forms a fundamental implementation strategy, where cryptographic operations are accessed through standardized interfaces that hide the details of underlying implementations. This approach enables applications to interact with cryptographic functions without needing to understand the specific algorithms being used, facilitating algorithm updates. Parameter configuration strategies that allow for dynamic adjustment of cryptographic parameters based on security requirements or performance considerations help maintain optimal protection while accommodating evolving standards.

The implementation strategies extend to various aspects of cryptographic system development. Algorithm negotiation protocols that enable communicating parties to agree on cryptographic algorithms based on their respective capabilities form an essential component of agile implementations. The development of cryptographic modules that can be updated independently helps ensure that changes to specific algorithms do not require complete system redesigns. The implementation of versioning mechanisms that maintain backward compatibility while supporting new algorithms helps prevent disruptions during cryptographic transitions. These implementation strategies combine to create systems that can adapt to changing security requirements while maintaining robust protection and operational continuity.

7.1.5 Testing and Validation

Comprehensive testing and validation form essential components of cryptographic agility by ensuring that systems can adapt to new algorithms while maintaining security guarantees. The testing process typically begins with functional verification that confirms cryptographic operations behave as expected under various conditions. This involves evaluating both the mathematical correctness of cryptographic implementations and their operational characteristics in different environments. Security validation forms a critical component of the testing process, evaluating resistance to both mathematical and implementation attacks across different algorithm configurations.

The validation process extends to various aspects of cryptographic system operation. Interoperability testing verifies that systems can communicate effectively when using different cryptographic algorithms or parameters, which becomes particularly important during transition periods. Performance profiling evaluates the operational impacts of different algorithm configurations to ensure that security updates do not introduce unacceptable overhead. Side-channel analysis examines implementations for physical vulnerabilities that might leak sensitive information through execution characteristics. These comprehensive testing and validation approaches help ensure that cryptographic systems can adapt to new algorithms while maintaining robust security guarantees and operational functionality.

7.1.6 Migration Strategies

Effective migration to quantum-resistant cryptographic systems requires strategies that leverage the principles of cryptographic agility to facilitate smooth transitions. These strategies typically begin with cryptographic inventory processes that identify all instances of classical algorithms within an organization's systems. The migration planning involves developing phased approaches that prioritize updates to the most vulnerable components while maintaining compatibility with existing implementations. Hybrid deployment strategies that support both classical and post-quantum algorithms simultaneously help ensure continuous protection during transition periods.

The migration strategies extend to various aspects of cryptographic system updates. Pilot implementations that test post-quantum algorithms in controlled environments before broader deployment help identify potential issues before full-scale updates. The development of rollback mechanisms that can revert to classical algorithms if significant problems arise with post-quantum implementations helps mitigate risks during transitions.

Performance optimization strategies that balance security requirements with operational efficiency help ensure that cryptographic updates do not disrupt system functionality. These migration strategies leverage the principles of cryptographic agility to facilitate effective transitions to quantum-resistant security while maintaining robust protection and operational continuity.

7.1.7 Challenges

The implementation of cryptographic agility principles presents several challenges that require careful consideration and strategic approaches. Algorithm selection involves evaluating multiple post-quantum candidates based on their security guarantees, performance characteristics, and compatibility with existing systems.

The complexity of many quantum-resistant algorithms makes their implementation particularly challenging, requiring specialized expertise that may not be readily available. The need to support multiple algorithms simultaneously can introduce compatibility issues that must be carefully managed to prevent disruptions during transition periods.

The challenges extend to various aspects of cryptographic system development. Performance optimization involves balancing security requirements with operational efficiency, where the computational demands of post-quantum algorithms often exceed those of classical primitives. The management of cryptographic parameters requires careful consideration of security implications when adjusting algorithm configurations. The development of standardized interfaces that can support multiple algorithms while maintaining security guarantees presents another significant challenge. These implementation challenges require comprehensive strategies that address both technical and operational considerations to ensure successful adoption of cryptographic agility principles.

7.2 INVENTORYING AND AUDITING CRYPTOGRAPHIC ASSETS

The inventorying and auditing of cryptographic assets form the essential foundation for transitioning to quantum-resistant security. As organizations prepare for the quantum computing era, comprehensive cryptographic asset management provides the necessary visibility and insights for developing effective migration strategies. The complex nature of modern digital ecosystems requires systematic approaches to identifying and evaluating cryptographic implementations, forming the basis for targeted updates to post-quantum algorithms.

The future of cryptographic security lies in the balanced development of asset management capabilities that address both technical and operational considerations. As quantum computing capabilities continue to advance, the cryptographic community's focus on comprehensive inventory and audit processes will remain crucial for enabling practical deployment of quantum-resistant security solutions.

The experiences gained through current asset management efforts will inform future developments, ensuring that organizations can effectively navigate the complex transition to quantum-safe cryptography while maintaining robust protection in an era of rapidly advancing technological capabilities.

The future of cryptographic asset management involves several promising directions that will shape the evolution of quantum-resistant security solutions. The integration of artificial intelligence techniques presents opportunities for more sophisticated cryptographic discovery and analysis. Machine learning approaches could help identify undocumented cryptographic implementations and predict potential migration challenges based on system characteristics. The development of automated inventory tools that can continuously monitor cryptographic assets and update inventories in real-time represents another promising direction.

The evolution of cryptographic asset management will likely focus on enhanced visualization capabilities that provide more intuitive representations of cryptographic ecosystems. The integration of predictive analytics that can forecast potential migration issues based on current system configurations may help organizations develop more effective transition strategies.

The development of standardized interfaces for cryptographic asset management tools could facilitate broader adoption and interoperability between different systems. These future directions will shape the development of cryptographic asset management, ensuring that organizations can effectively navigate the complex transition to quantum-resistant security while maintaining robust protection in an evolving technological landscape.

7.2.1 Scope

Comprehensive cryptographic inventory processes involve systematic identification of all cryptographic assets across an organization's digital infrastructure. This scope extends beyond obvious cryptographic implementations to include embedded systems, third-party integrations, and legacy components that may rely on outdated security mechanisms. The inventory process must account for various cryptographic functions, including encryption algorithms, digital signature schemes, key exchange protocols, and authentication mechanisms. Each of these components requires careful evaluation to determine its role in the overall security architecture and its potential impact on migration strategies.

The inventory scope also encompasses different layers of digital infrastructure, from application-level cryptographic functions to network protocols and hardware security modules. This comprehensive approach helps ensure that no cryptographic implementation is overlooked during the transition process. The inventory must consider both active cryptographic operations and stored cryptographic materials, including keys, certificates, and configuration parameters. This holistic view of cryptographic assets provides the necessary foundation for developing effective migration strategies that address all aspects of an organization's security infrastructure.

7.2.2 Methodologies for Cryptographic Asset Discovery

Effective cryptographic asset discovery employs multiple methodologies to identify implementations across diverse digital environments. Static analysis techniques examine code repositories, configuration files, and system documentation to catalog cryptographic algorithm usage. These methods can identify explicit cryptographic function calls and configuration parameters that specify security algorithms. Dynamic analysis approaches monitor runtime behavior to detect cryptographic operations in executing systems, helping identify implementations that might not be apparent through static examination alone. The combination of these techniques provides a more complete picture of cryptographic usage across complex digital ecosystems.

Advanced discovery methodologies incorporate machine learning techniques that can identify patterns indicative of cryptographic implementations. These approaches analyze system behavior to detect cryptographic operations that might not be explicitly documented. Network traffic analysis examines communication patterns to identify cryptographic protocols in use. The integration of these advanced techniques with traditional discovery methods helps create comprehensive inventories that form the basis for effective migration planning. The discovery process must be iterative, as new cryptographic implementations may be introduced during the transition period, requiring ongoing inventory updates.

7.2.3 Classification and Prioritization

The classification of cryptographic assets involves evaluating each implementation based on its security criticality and migration complexity. This process considers the sensitivity of data protected by each cryptographic function, the potential impact of algorithm updates on system operations, and the feasibility of migration to post-quantum alternatives. High-priority assets typically include those protecting sensitive information or supporting critical infrastructure, where the consequences of cryptographic vulnerabilities would be most severe. The classification process also evaluates the interdependencies between different cryptographic components, as updates to one system may require corresponding changes in others.

The prioritization process involves developing a migration roadmap that sequences cryptographic updates based on their criticality and complexity. This roadmap typically begins with the most vulnerable systems that would benefit most from immediate updates to post-quantum algorithms. The prioritization also considers the operational impacts of cryptographic changes, sequencing updates to minimize disruptions to business continuity. The classification and prioritization processes help organizations allocate resources effectively during the transition, ensuring that the most critical systems receive attention first while developing comprehensive plans for complete migration.

7.2.4 Auditing Cryptographic Implementations

The auditing process evaluates cryptographic implementations to assess their security characteristics and potential vulnerabilities. This involves examining algorithm configurations to verify they meet current security standards and identifying potential weaknesses that might be exploited by quantum attacks. The audit process also evaluates implementation characteristics that could introduce vulnerabilities, such as side-channel attack vectors or improper parameter usage. These evaluations help determine the urgency of updates to post-quantum algorithms and identify potential challenges that might arise during migration.

The auditing process extends to evaluating the operational impacts of cryptographic implementations. This includes assessing performance characteristics that might be affected by algorithm updates and identifying compatibility issues that could arise during transition periods. The audit also examines key management practices to ensure they can accommodate the larger key sizes characteristic of many post-quantum algorithms. These comprehensive evaluations provide the necessary insights for developing effective migration strategies that address both security requirements and operational constraints.

7.2.5 Interdependency Analysis in Cryptographic Systems

The analysis of interdependencies between cryptographic components forms a critical aspect of inventorying and auditing processes. Modern digital ecosystems often involve complex relationships between different systems and services, where updates to one cryptographic implementation may affect others.

This interdependency analysis examines communication protocols that rely on specific cryptographic algorithms, authentication mechanisms that depend on particular security functions, and data formats that incorporate cryptographic protections. The analysis helps identify potential compatibility issues that might arise during migration and informs the development of strategies to address these challenges.

The interdependency analysis also considers the temporal aspects of cryptographic transitions. Some systems may need to maintain compatibility with classical algorithms while others transition to post-quantum alternatives, requiring careful coordination of updates. The analysis evaluates the potential for hybrid implementations that can support multiple algorithms simultaneously to facilitate smooth transitions. These considerations help develop migration strategies that maintain operational continuity while updating cryptographic implementations to quantum-resistant alternatives.

7.2.6 Tooling for Cryptographic Inventory and Audit

Specialized tools for cryptographic inventory and audit provide essential capabilities for managing the transition to post-quantum security. These tools typically incorporate scanning capabilities that can identify cryptographic implementations across diverse digital environments. The scanning functions often combine static and dynamic analysis techniques to create comprehensive inventories of cryptographic assets. The tools also provide visualization features that help organizations understand the relationships between different cryptographic components and the potential impacts of algorithm updates.

Advanced inventory tools incorporate machine learning techniques that can identify cryptographic usage patterns that might indicate undocumented implementations. These tools often include reporting capabilities that generate detailed assessments of cryptographic assets, including their security characteristics and migration complexities. The audit functions typically evaluate implementations against security standards and identify potential vulnerabilities that might be exploited by quantum attacks. These comprehensive tooling capabilities help organizations develop effective migration strategies that address both security requirements and operational constraints.

7.2.7 Developing Migration Strategies Based on Inventory Results

The results of cryptographic inventory and audit processes form the foundation for developing effective migration strategies. These strategies typically begin with pilot implementations that test post-quantum algorithms in controlled environments before broader deployment. The migration planning involves developing phased approaches that prioritize updates to the most vulnerable systems while maintaining compatibility with existing implementations. The strategies also consider the operational impacts of cryptographic changes, sequencing updates to minimize disruptions to business continuity.

The migration strategies incorporate contingency plans that address potential challenges that might arise during the transition. These plans typically include rollback mechanisms that can revert to classical algorithms if significant issues emerge with post-quantum implementations. The strategies also consider performance optimization techniques that can help mitigate the computational impacts of algorithm updates. These comprehensive migration approaches help organizations navigate the complex process of cryptographic transitions while maintaining robust security guarantees and operational functionality.

7.2.8 Challenges

The process of inventorying and migrating cryptographic assets presents several challenges that require careful consideration and strategic approaches. The complexity of modern digital ecosystems makes comprehensive cryptographic discovery particularly challenging, as implementations may be embedded in various layers of system architecture.

The diversity of cryptographic algorithms and their implementations complicates the inventory process, requiring specialized knowledge to identify and evaluate different security functions. The interdependencies between cryptographic components introduce additional complexity, as updates to one system may affect others in unpredictable ways.

The migration challenges extend to various aspects of cryptographic transitions. The computational requirements of many post-quantum algorithms often exceed those of classical primitives, potentially introducing performance bottlenecks in resource-constrained environments. The larger key sizes characteristic of quantum-resistant schemes can strain memory resources and communication bandwidth, requiring careful optimization. The need to maintain compatibility with existing systems during transition periods introduces additional complexity, as hybrid implementations must support multiple algorithms simultaneously.

These challenges require comprehensive strategies that address both technical and operational considerations to ensure successful migration to quantum-resistant cryptography.

7.3 UPDATING TLS, VPNS, AND PKI INFRASTRUCTURES

The update of fundamental security protocols to quantum-resistant alternatives represents a critical component of the transition to a quantum-safe world. As organizations prepare for the quantum computing era, comprehensive strategies for modernizing TLS, VPN, and PKI infrastructures provide the necessary framework for maintaining secure communications. The complex nature of these updates requires careful planning and execution to ensure that security improvements do not come at the expense of operational continuity.

The future of digital security lies in the balanced development of protocol modernization strategies that address both technical and operational considerations. As quantum computing capabilities continue to advance, the cryptographic community's focus on updating fundamental security protocols will remain crucial for enabling practical deployment of quantum-resistant security solutions.

The experiences gained through current modernization efforts will inform future developments, ensuring that digital communications can adapt to meet the challenges of advancing computational capabilities while maintaining robust protection in an era of rapidly evolving technological threats.

The future development of quantum-resistant security protocols involves several promising directions that will shape the evolution of digital communications. The integration of artificial intelligence techniques presents opportunities for more sophisticated cryptographic analysis and protocol optimization. Machine learning approaches could help identify optimal algorithm configurations based on specific security requirements and operational constraints. The development of adaptive protocols that can adjust their cryptographic parameters based on available resources or threat environments may provide enhanced protection while maintaining performance characteristics.

The evolution of security protocols will likely focus on enhanced negotiation capabilities that can support an increasingly diverse set of cryptographic algorithms. The development of standardized interfaces for cryptographic operations could facilitate broader adoption and interoperability between different systems. The integration of performance monitoring and adaptive optimization features will help ensure that security protocols can maintain operational efficiency while accommodating evolving security requirements. These future directions will shape the development of quantum-resistant security protocols, ensuring that digital communications can adapt to meet the challenges of advancing computational capabilities while maintaining robust protection in an evolving technological landscape.

7.3.1 Challenges

The process of updating fundamental security protocols to quantum-resistant alternatives presents significant challenges that require careful navigation. The interconnected nature of modern digital ecosystems means that updates to one component may affect others in unpredictable ways, necessitating comprehensive planning and coordination.

TLS, VPN, and PKI infrastructures often involve complex dependencies between different systems and services, where cryptographic updates must be carefully synchronized to prevent compatibility issues or security vulnerabilities. The pervasive use of these protocols across diverse applications and environments further complicates the update process, requiring strategies that can accommodate varying operational requirements and constraints.

Another significant challenge stems from the need to maintain backward compatibility during transition periods. Many systems will need to support both classical and post-quantum algorithms simultaneously to ensure uninterrupted communications while gradually adopting quantum-resistant alternatives. This requirement introduces additional complexity in protocol design and implementation, as systems must negotiate cryptographic parameters and handle different algorithm configurations. The computational requirements of many post-quantum algorithms often exceed those of classical primitives, potentially introducing performance bottlenecks that must be carefully managed to prevent disruptions to critical operations.

7.3.2 Updating Transport Layer Security Protocols

The transition of Transport Layer Security protocols to quantum-resistant alternatives requires careful consideration of both security requirements and operational constraints. TLS forms the foundation of secure communications across the internet, protecting everything from web browsing to financial transactions. Updating this fundamental protocol involves not just the adoption of new cryptographic algorithms but also the development of negotiation mechanisms that can handle both classical and post-quantum algorithms during transition periods. The update process must consider the performance impacts of quantum-resistant algorithms, which often require more computational resources and larger key sizes than classical primitives.

The TLS update strategy typically begins with the identification of all protocol implementations across an organization's digital infrastructure. This inventory process helps determine the scope of updates required and the potential impacts on system operations. The transition plan often involves phased implementations that prioritize updates to the most vulnerable systems while maintaining compatibility with existing configurations. Hybrid implementations that can support multiple algorithms simultaneously help ensure continuous protection during the transition period. The update process must also consider the potential for downgrade attacks, where adversaries might attempt to force systems to use weaker classical algorithms, requiring careful design of negotiation protocols to prevent such vulnerabilities.

7.3.3 Transitioning Virtual Private Network Infrastructures

The migration of Virtual Private Network infrastructures to quantum-resistant security presents unique challenges stemming from their role in securing remote communications. VPNs often protect sensitive organizational communications, making their timely update particularly crucial in the transition to post-quantum cryptography.

The update process involves not just the adoption of new encryption algorithms but also the development of key exchange mechanisms that can establish secure sessions using quantum-resistant primitives. The performance characteristics of these new algorithms become particularly important for VPN implementations, where latency and throughput requirements can significantly impact user experience and operational efficiency.

The VPN transition strategy typically begins with a comprehensive assessment of existing implementations to identify all cryptographic components requiring updates. This evaluation helps determine the potential impacts of algorithm changes on network performance and compatibility with client devices. The migration plan often involves pilot implementations that test post-quantum algorithms in controlled environments before broader deployment.

These tests help identify potential performance bottlenecks or compatibility issues that must be addressed before full-scale updates. The transition process must also consider the potential for man-in-the-middle attacks during negotiation phases, requiring careful design of key exchange protocols to maintain security throughout the migration period.

7.3.4 Modernizing Public Key Infrastructure Systems

The update of Public Key Infrastructure systems to quantum-resistant alternatives represents one of the most complex aspects of the cryptographic transition. PKIs form the foundation of digital trust, providing the authentication and encryption mechanisms that secure modern digital ecosystems. The migration process involves not just the adoption of new signature and encryption algorithms but also the development of certificate formats that can accommodate larger post-quantum keys and parameters. The hierarchical nature of PKI systems introduces additional complexity, as updates must be carefully coordinated across certificate authorities, registration authorities, and relying parties to prevent disruptions to the chain of trust.

The PKI modernization strategy typically begins with an inventory of all certificate authorities and relying party implementations to determine the scope of required updates. This assessment helps identify potential compatibility issues that might arise when some components transition to post-quantum algorithms while others remain on classical schemes.

The migration plan often involves developing hybrid certificate formats that can support both algorithm types during transition periods. The update process must also consider the potential for certificate forgery attacks that might exploit vulnerabilities in the migration process, requiring careful design of signature schemes and validation procedures to maintain security throughout the transition.

7.3.5 Interoperability Considerations during Transition

The maintenance of interoperability between systems using different cryptographic algorithms forms a critical aspect of the transition to quantum-resistant security. As organizations update their TLS, VPN, and PKI infrastructures at varying paces, the need to support multiple algorithm types simultaneously becomes increasingly important. This requirement introduces additional complexity in protocol design and implementation, as systems must negotiate cryptographic parameters and handle different algorithm configurations without compromising security or performance. The interoperability challenges become particularly pronounced in heterogeneous environments where different organizations may progress through migration timelines at different rates.

The development of cryptographic negotiation protocols that can agree on algorithm parameters based on the capabilities of communicating parties helps address these interoperability challenges. These protocols must be carefully designed to prevent downgrade attacks while maintaining compatibility with existing implementations. The use of algorithm agility features that can switch between different cryptographic primitives as needed provides additional flexibility during transition periods.

The interoperability considerations also extend to certificate formats and key exchange mechanisms, which must be designed to accommodate varying cryptographic capabilities while maintaining security guarantees. These comprehensive approaches help ensure that systems can communicate securely throughout the migration process despite differing cryptographic implementations.

7.3.6 Performance Optimization Strategies

The performance impacts of post-quantum cryptographic algorithms on fundamental security protocols require careful optimization strategies to maintain operational efficiency. Many quantum-resistant algorithms involve more complex mathematical operations and larger key sizes than classical primitives, potentially introducing significant computational overhead.

The performance characteristics of these new algorithms become particularly important for TLS and VPN implementations, where latency and throughput requirements can significantly affect user experience and system functionality. The optimization strategies must balance security requirements with operational constraints to ensure that cryptographic updates do not disrupt critical operations.

The performance optimization process typically begins with algorithm selection that considers both security guarantees and computational requirements. This involves evaluating different post-quantum candidates based on their performance characteristics in specific operational environments. The optimization strategies often incorporate hardware acceleration techniques that can significantly improve the efficiency of cryptographic operations.

These might include specialized processors for post-quantum algorithms or cryptographic co-processors that can handle complex mathematical operations. The optimization process also considers protocol design decisions that can minimize the performance impacts of algorithm updates, such as reducing the number of cryptographic operations required for secure communications.

7.3.7 Testing and Validation Methodologies

Comprehensive testing and validation form essential components of the transition to quantum-resistant security protocols. The testing process typically begins with functional verification that confirms cryptographic operations behave as expected under various conditions. This involves evaluating both the mathematical correctness of post-quantum implementations and their operational characteristics in different environments. Security validation forms a critical component of the testing process, assessing resistance to both mathematical and implementation attacks across different algorithm configurations. The validation process must consider the unique characteristics of TLS, VPN, and PKI infrastructures, where different protocols present specific security requirements and potential vulnerabilities.

The testing methodologies extend to various aspects of protocol operation. Interoperability testing verifies that systems can communicate effectively when using different cryptographic algorithms or parameters, which becomes particularly important during transition periods. Performance profiling evaluates the operational impacts of different algorithm configurations to ensure that security updates do not introduce unacceptable overhead. Side-channel analysis examines implementations for physical vulnerabilities that might leak sensitive information through execution characteristics. These comprehensive testing approaches help ensure that updated security protocols maintain robust protection while functioning correctly in operational environments.

7.3.8 Migration Strategies for Fundamental Security Protocols

Effective migration to quantum-resistant security protocols requires strategies that leverage the principles of cryptographic agility to facilitate smooth transitions. These strategies typically begin with pilot implementations that test post-quantum algorithms in controlled environments before broader deployment. The migration planning involves developing phased approaches that prioritize updates to the most vulnerable systems while maintaining compatibility with existing implementations. Hybrid implementations that can support both classical and post-quantum algorithms simultaneously help ensure continuous protection during the transition period. The migration strategies must also consider the potential for rollback mechanisms that can revert to classical algorithms if significant issues emerge with post-quantum implementations.

The migration strategies extend to various aspects of protocol operation. The development of algorithm negotiation protocols that can agree on cryptographic parameters based on the capabilities of communicating parties helps address interoperability challenges. The use of versioning mechanisms that maintain backward compatibility while supporting new algorithms provides additional flexibility during transition periods. The migration strategies also consider performance optimization techniques that can help mitigate the computational impacts of algorithm updates. These comprehensive approaches help organizations navigate the complex process of protocol transitions while maintaining robust security guarantees and operational functionality.

7.4 SECURE SOFTWARE DEVELOPMENT WITH PQC

The development of secure software with post-quantum cryptography represents a fundamental component of the transition to a quantum-safe world. As organizations prepare for the quantum computing era, comprehensive strategies for incorporating quantum-resistant algorithms into software applications provide the necessary framework for maintaining digital security. The complex nature of this transition requires careful planning and execution to ensure that security improvements do not come at the expense of operational functionality.

The future of quantum-safe software development lies in the balanced evolution of security and performance characteristics that address both current and emerging threats. As quantum computing capabilities continue to advance, the software development community's focus on secure implementation practices will remain crucial for enabling practical deployment of quantum-resistant security solutions. The experiences gained through current development efforts will inform future advancements, ensuring that software applications can adapt to meet the challenges of evolving computational capabilities while maintaining robust protection in the quantum era. The principles and practices of secure software development with post-quantum cryptography provide the foundation for building a quantum-safe digital future.

The future of secure software development with post-quantum cryptography involves several promising directions that will shape the evolution of quantum-resistant security solutions. The integration of artificial intelligence techniques presents opportunities for more sophisticated cryptographic analysis and implementation optimization. Machine learning approaches could help identify optimal algorithm configurations based on specific security requirements and operational constraints. The development of automated testing frameworks that can evaluate post-quantum implementations for potential vulnerabilities may provide enhanced security validation capabilities.

The evolution of quantum-safe software development will likely focus on enhanced abstraction capabilities that further simplify the adoption of new algorithms. The development of standardized interfaces that can support an increasingly diverse set of cryptographic primitives may facilitate broader adoption and interoperability between different systems. The integration of performance monitoring and adaptive optimization features will help ensure that software can maintain operational efficiency while accommodating evolving security requirements. These future directions will shape the development of quantum-safe software, ensuring that applications can adapt to meet the challenges of advancing computational capabilities while maintaining robust protection in an evolving technological landscape.

7.4.1 Fundamental Principles

The principles of secure software development with post-quantum cryptography build upon established security practices while incorporating new considerations specific to quantum-resistant algorithms. Cryptographic agility emerges as a fundamental principle, where software is designed to support multiple cryptographic algorithms through standardized interfaces. This approach enables the replacement of specific cryptographic primitives without requiring complete application redesigns, facilitating transitions to new algorithms as standards evolve. Parameter flexibility represents another key principle, allowing for adjustments to cryptographic parameters as security requirements change or new vulnerabilities are discovered. These principles help ensure that software can adapt to the evolving security landscape while maintaining robust protection. The principles extend to various aspects of software development. Algorithm abstraction forms an essential component, where cryptographic operations are accessed through standardized interfaces that hide the details of underlying implementations.

This approach enables applications to interact with cryptographic functions without needing to understand the specific algorithms being used, facilitating transitions to new primitives. Security validation becomes particularly important for post-quantum implementations, where new cryptographic primitives may introduce different implementation vulnerabilities than classical algorithms. These principles combine to create software that can adapt to changing cryptographic standards while maintaining security guarantees and operational functionality.

7.4.2 Secure Implementation Practices

The secure implementation of post-quantum cryptographic algorithms requires specialized practices that address the unique characteristics of these new primitives. Constant-time programming techniques form a fundamental practice, where cryptographic operations are designed to complete in uniform time regardless of secret values. This approach helps prevent timing attacks that could exploit variations in execution time to infer sensitive information. Memory access patterns must be carefully structured to avoid cache vulnerabilities that might leak information about secret values through timing variations. These implementation practices become particularly important for post-quantum algorithms, where complex mathematical operations may introduce new side-channel vulnerabilities if not properly addressed.

The secure implementation practices extend to various aspects of software development. Hardware acceleration techniques can significantly improve the performance of post-quantum algorithms while maintaining security guarantees. These techniques must be carefully designed to prevent physical attacks that could exploit hardware characteristics to compromise cryptographic security. The implementation of post-quantum algorithms must also consider the larger key sizes and parameter requirements characteristic of these new primitives, which can introduce additional implementation challenges. These secure implementation practices help ensure that post-quantum cryptographic software maintains robust protection against both mathematical and physical attacks.

7.4.3 Integration Strategies

The integration of post-quantum cryptographic algorithms into software applications requires careful strategies that address both security and compatibility considerations. Hybrid integration approaches form a fundamental strategy, where software can support both classical and post-quantum algorithms simultaneously. This enables gradual transitions to new cryptographic primitives while maintaining compatibility with existing systems. The integration strategies must consider the performance impacts of post-quantum algorithms, which often require more computational resources than classical primitives. These impacts can be particularly significant for resource-constrained environments, where the computational demands of quantum-resistant algorithms may strain limited processing capabilities.

The integration strategies extend to various aspects of software architecture. Algorithm negotiation protocols enable communicating parties to agree on cryptographic algorithms based on their respective capabilities, facilitating interoperability during transition periods. The development of cryptographic modules that can be updated independently helps ensure that changes to specific algorithms do not require complete application redesigns. The integration of post-quantum algorithms must also consider the potential for algorithm downgrade attacks, where adversaries might attempt to force systems to use weaker classical algorithms. These integration strategies help ensure that software can incorporate post-quantum cryptographic primitives while maintaining security guarantees and operational functionality.

7.4.4 Testing and Validation Methodologies

Comprehensive testing and validation form essential components of secure software development with post-quantum cryptography. The testing process typically begins with functional verification that confirms cryptographic operations behave as expected under various conditions. This involves evaluating both the mathematical correctness of post-quantum implementations and their operational characteristics in different environments. Security validation forms a critical component of the testing process, assessing resistance to both mathematical and implementation attacks across different algorithm configurations. The validation process must consider the unique characteristics of post-quantum algorithms, where new cryptographic primitives may introduce different vulnerabilities than classical schemes.

The testing methodologies extend to various aspects of software operation. Side-channel analysis examines implementations for physical vulnerabilities that might leak sensitive information through execution characteristics. This analysis becomes particularly important for post-quantum implementations, where complex mathematical operations may introduce new attack vectors. Performance profiling evaluates the operational impacts of different algorithm configurations to ensure that security updates do not introduce unacceptable overhead. Interoperability testing verifies that software can communicate effectively when using different cryptographic algorithms or parameters, which becomes particularly important during transition periods. These comprehensive testing approaches help ensure that post-quantum cryptographic software maintains robust protection while functioning correctly in operational environments.

7.4.5 Performance Optimization Techniques

The performance optimization of post-quantum cryptographic software involves specialized techniques that address the unique characteristics of these new algorithms. Algorithm selection forms a fundamental optimization technique, where developers choose post-quantum primitives based on their security guarantees and performance characteristics in specific operational environments. Parameter optimization provides another technique, where cryptographic parameters are adjusted to balance security requirements with computational efficiency. These optimization techniques must consider the larger key sizes and more complex operations characteristic of post-quantum algorithms, which can introduce significant performance overhead if not properly managed.

The optimization techniques extend to various aspects of software implementation. Hardware acceleration can significantly improve the performance of post-quantum algorithms through specialized processors or cryptographic co-processors. These techniques must be carefully designed to maintain security guarantees while enhancing computational efficiency. The optimization of memory access patterns can help mitigate the performance impacts of larger key sizes and expanded parameter requirements.

The development of efficient data structures and algorithms that minimize computational overhead while maintaining security becomes particularly important for resource-constrained environments. These performance optimization techniques help ensure that post-quantum cryptographic software can be effectively deployed in diverse operational scenarios.

7.4.6 Migration Strategies for Software Development

The migration to post-quantum cryptographic software requires strategies that leverage the principles of cryptographic agility to facilitate smooth transitions. These strategies typically begin with pilot implementations that test post-quantum algorithms in controlled environments before broader deployment. The migration planning involves developing phased approaches that prioritize updates to the most vulnerable software components while maintaining compatibility with existing implementations.

Hybrid implementations that can support both classical and post-quantum algorithms simultaneously help ensure continuous protection during the transition period. The migration strategies must also consider the potential for rollback mechanisms that can revert to classical algorithms if significant issues emerge with post-quantum implementations.

The migration strategies extend to various aspects of software development. The development of algorithm negotiation capabilities that can agree on cryptographic parameters based on the capabilities of communicating parties helps address interoperability challenges. The use of versioning mechanisms that maintain backward compatibility while supporting new algorithms provides additional flexibility during transition periods. The migration strategies also consider performance optimization techniques that can help mitigate the computational impacts of algorithm updates. These comprehensive approaches help organizations navigate the complex process of software transitions while maintaining robust security guarantees and operational functionality.

7.4.7 Challenges

The development of quantum-safe software presents several challenges that require careful consideration and strategic approaches. The complexity of post-quantum algorithms makes their secure implementation particularly challenging, as these new primitives often introduce different vulnerabilities than classical cryptographic schemes.

The larger key sizes and more complex operations characteristic of quantum-resistant algorithms can strain computational resources, potentially introducing performance bottlenecks in resource-constrained environments. The need to maintain compatibility with existing systems during transition periods introduces additional complexity, as software must support multiple algorithms simultaneously without compromising security or functionality.

The challenges extend to various aspects of software development. The selection of appropriate post-quantum algorithms requires careful evaluation of their security guarantees and performance characteristics in specific operational environments. The implementation of these new primitives must address potential side-channel vulnerabilities that could compromise security through physical characteristics of execution. The management of cryptographic parameters becomes particularly important for post-quantum algorithms, where different configurations can significantly impact both security and performance. These challenges require comprehensive strategies that address both technical and operational considerations to ensure successful development of quantum-safe software.

7.5 RISK MITIGATION IN HYBRID DEPLOYMENT

The implementation of hybrid cryptographic deployments represents a critical strategy for transitioning to quantum-resistant security solutions. As organizations prepare for the quantum computing era, comprehensive risk mitigation strategies provide the necessary framework for maintaining security while adapting to new cryptographic standards. The complex nature of these hybrid deployments requires careful planning and execution to ensure that security improvements do not come at the expense of operational continuity.

The future of quantum-safe security lies in the balanced development of hybrid deployment strategies that address both technical and operational considerations. As quantum computing capabilities continue to advance, the cryptographic community's focus on risk mitigation in hybrid environments will remain crucial for enabling practical deployment of quantum-resistant security solutions. The experiences gained through current hybrid deployment efforts will inform future developments, ensuring that organizations can effectively navigate the complex transition to quantum-safe cryptography while maintaining robust protection in an era of rapidly advancing technological capabilities.

The future development of hybrid cryptographic systems involves several promising directions that will shape the evolution of quantum-resistant security solutions. The integration of artificial intelligence techniques presents opportunities for more sophisticated algorithm selection and negotiation strategies. Machine learning approaches could help identify optimal cryptographic configurations based on specific security requirements and operational constraints. The development of automated testing frameworks that can evaluate hybrid implementations for potential vulnerabilities may provide enhanced security validation capabilities.

The evolution of hybrid cryptographic systems will likely focus on enhanced negotiation capabilities that can support an increasingly diverse set of cryptographic algorithms. The development of standardized interfaces that can facilitate interoperability between different systems may help ensure continuous operation during transition periods. The integration of performance monitoring and adaptive optimization features will help maintain operational efficiency while accommodating evolving security requirements. These future directions will shape the development of hybrid cryptographic systems, ensuring that security infrastructures can adapt to meet the challenges of advancing computational capabilities while maintaining robust protection in an evolving technological landscape.

CHAPTER

8

Post-Quantum Cryptography in Applications

8.1 SECURE MESSAGING AND EMAIL (E.G., SIGNAL, PROTONMAIL)

The transition to post-quantum cryptography is not a matter of if but when. Secure messaging and email services must proactively adopt quantum-resistant algorithms to protect against future threats. Platforms like Signal and ProtonMail are leading the way by experimenting with hybrid cryptographic systems and collaborating with researchers to refine PQC implementations. While challenges remain, the combined efforts of the cybersecurity community, standardization bodies, and technology companies will ensure that secure communication remains viable in the quantum era. Users, businesses, and governments must stay informed and prepared to embrace these advancements, safeguarding privacy and security in an increasingly interconnected world. The shift to post-quantum cryptography represents a necessary evolution in digital security, ensuring that encrypted communications remain confidential and tamper-proof for decades to come.

The future of secure messaging and email lies in the seamless integration of post-quantum cryptography without sacrificing usability or performance. Researchers and developers are working on optimizing PQC algorithms to reduce their computational and bandwidth requirements. Additionally, advancements in hardware acceleration, such as quantum-resistant cryptographic chips, could help mitigate performance bottlenecks. Standardization bodies like NIST are expected to finalize their PQC standards in the coming years, providing a clear roadmap for implementation. Meanwhile, user education will play a crucial role in ensuring that individuals understand the importance of upgrading to quantum-secure communication tools.

8.1.1 Current State of Secure Messaging and Email Services

Secure messaging applications such as Signal, WhatsApp, and Telegram, along with encrypted email services like ProtonMail and Tutanota, rely on end-to-end encryption (E2EE) to protect user communications. Signal, for instance, uses the Signal Protocol, which combines the Double Ratchet Algorithm with elliptic-curve Diffie-Hellman (ECDH) for key exchange. Similarly, ProtonMail employs PGP (Pretty Good Privacy) with RSA and AES for email encryption. While these methods are currently secure against classical computing attacks, they are not inherently resistant to quantum algorithms such as Shor's algorithm, which can break traditional public-key cryptography. This vulnerability necessitates the integration of post-quantum cryptographic techniques into these platforms.

8.1.2 Threats Posed by Quantum Computing to Secure Communication

Quantum computers, once fully realized, will be capable of solving certain mathematical problems exponentially faster than classical computers. Shor's algorithm, for example, can factor large integers and compute discrete logarithms—tasks that underpin the security of RSA and ECC—in polynomial time. This capability means that an attacker with a sufficiently powerful quantum computer could decrypt past communications if they have recorded encrypted data. Secure messaging and email services must therefore transition to quantum-resistant algorithms to prevent such retroactive decryption. The threat is not merely theoretical; governments and corporations are already storing encrypted data in anticipation of future quantum decryption, a strategy known as "harvest now, decrypt later."

8.1.3 Post-Quantum Cryptography in Secure Messaging

Several secure messaging platforms are beginning to explore post-quantum cryptography to future-proof their encryption protocols. Signal, a leader in privacy-focused messaging, has introduced PQXDH (Post-Quantum Extended Diffie-Hellman), a hybrid key exchange mechanism that combines classical ECDH with a post-quantum key encapsulation method (KEM). This approach ensures that even if one of the cryptographic systems is compromised, the other remains secure. Similarly, other messaging apps are expected to follow suit by integrating lattice-based, hash-based, or code-based cryptographic schemes, which are currently the most promising candidates for PQC standardization by NIST.

8.1.4 Post-Quantum Cryptography in Encrypted Email Services

Email encryption services like ProtonMail and Tutanota also face the challenge of transitioning to quantum-resistant cryptography. ProtonMail has announced plans to integrate post-quantum secure algorithms into its encryption framework, ensuring that emails remain protected against quantum attacks. One approach involves combining traditional PGP encryption with post-quantum algorithms, creating a hybrid system that maintains compatibility while enhancing security. Additionally, efforts are underway to develop quantum-resistant digital signatures, which are essential for verifying the authenticity of emails. The migration to PQC in email services must be carefully managed to avoid disruptions while ensuring long-term security.

8.1.5 Challenges in Adopting Post-Quantum Cryptography

Despite the clear advantages of post-quantum cryptography, several challenges hinder its widespread adoption in secure messaging and email services. One major obstacle is the increased computational overhead associated with PQC algorithms. Many quantum-resistant schemes, such as lattice-based cryptography, require larger key sizes and more processing power, which could impact performance on mobile devices and low-power systems. Another challenge is interoperability—ensuring that new PQC-enabled applications remain compatible with older systems during the transition period. Additionally, standardization efforts by NIST are still ongoing, meaning that some algorithms may change or be replaced as research progresses.

8.1.6 Role of Hybrid Cryptography in Transition

To address these challenges, many secure communication platforms are adopting hybrid cryptographic systems that combine classical and post-quantum algorithms. This approach provides a safety net, ensuring that even if one cryptographic method is broken, the other remains secure. For example, a hybrid key exchange might use both ECDH and a post-quantum KEM, requiring an attacker to break both to compromise the communication. Hybrid systems also allow for a smoother transition, as they maintain backward compatibility with existing infrastructure while gradually introducing quantum-resistant mechanisms. This strategy is particularly important for widely used services like Signal and ProtonMail, where sudden cryptographic changes could disrupt user experience.

8.2 SECURE WEB COMMUNICATIONS (TLS 1.3 + PQC)

Secure web communication is the backbone of the digital economy and modern society. TLS 1.3 has already raised the bar for privacy and performance. However, with the looming threat of quantum computing, its reliance on classical public-key cryptography poses a significant vulnerability. Post-quantum cryptography offers a promising solution to this problem. The integration of PQC into TLS 1.3 represents a proactive step towards safeguarding digital communications against future threats. While there are technical and logistical challenges to overcome, the momentum is clearly building. With ongoing research, international collaboration, and forward-thinking implementation strategies, the vision of a quantum-secure internet is rapidly becoming a reality.

The journey toward fully secure quantum-resistant web communications involves several challenges. These include algorithm selection, interoperability, performance tuning, and security validation. However, the opportunities are immense. By adopting PQC early, organizations can future-proof their digital infrastructure against emerging threats. Educational institutions, developers, and security professionals must stay informed about advancements in PQC and contribute to its safe and widespread adoption. The integration of PQC into TLS 1.3 is not merely a technical update; it represents a paradigm shift in how the digital world approaches security in the face of transformative technological change.

8.2.1 TLS 1.3 and Its Role in Web Security

TLS 1.3 is the latest version of the TLS protocol and was officially published by the Internet Engineering Task Force (IETF) in 2018. It is designed to provide privacy and data integrity between two communicating applications. The most common usage of TLS is in securing the HTTPS protocol used by web browsers. TLS 1.3 offers significant improvements over its predecessors by removing outdated cryptographic algorithms, reducing handshake latency, and simplifying the protocol structure. It eliminates support for algorithms that have known weaknesses, such as RSA key exchange and SHA-1, and focuses on strong, modern alternatives like elliptic curve cryptography (ECC) and authenticated encryption with associated data (AEAD) ciphers. These enhancements contribute to both better performance and stronger security.

8.2.2 Quantum Threat to Cryptographic Protocols

Quantum computers, though still in their developmental stages, have the potential to break many cryptographic algorithms that are currently in widespread use. The most notable example is Shor's algorithm, which can efficiently factor large integers and compute discrete logarithms. These capabilities directly threaten RSA, DSA, and ECC—the foundational algorithms used in TLS for key exchange and digital signatures. In a quantum-enabled future, an attacker with a sufficiently powerful quantum computer could decrypt previously secure communications or forge digital signatures, compromising the integrity of sensitive information. Although quantum computers capable of executing Shor's algorithm on a large scale are not yet available, the possibility of their emergence has prompted the cryptographic community to prepare in advance. This preparation involves developing quantum-resistant algorithms that can be seamlessly integrated into existing security protocols.

8.2.3 Post-Quantum Cryptography and Its Foundations

Post-Quantum Cryptography refers to cryptographic algorithms that are believed to be secure against both classical and quantum computational attacks. Unlike quantum cryptography, which relies on the principles of quantum mechanics, PQC is based on hard mathematical problems that even quantum computers cannot solve efficiently. Examples include lattice-based cryptography, code-based cryptography, multivariate polynomial cryptography, and hash-based signatures. These schemes form the basis of candidate algorithms submitted to the NIST Post-Quantum Cryptography Standardization project. NIST's project has spurred the development and evaluation of new cryptographic standards that will eventually replace or supplement existing public-key systems. Several algorithms from this effort, such as Kyber for key encapsulation and Dilithium for digital signatures, have emerged as leading contenders for standardization.

8.2.4 Integrating PQC into TLS 1.3

The integration of post-quantum cryptographic algorithms into TLS 1.3 is both a technical and strategic challenge. TLS 1.3 relies heavily on asymmetric cryptography during its handshake phase for key exchange and authentication. Replacing or augmenting these operations with PQC algorithms requires modifications to the handshake protocol. To facilitate this, hybrid key exchange mechanisms have been proposed. In a hybrid approach, a traditional key exchange method like ECDHE is used in conjunction with a post-quantum key exchange scheme. The resulting shared secret is derived from both methods, ensuring that the connection remains secure even if one method is compromised. This approach offers a practical transition strategy while the PQC algorithms are still being evaluated and optimized for real-world use. Open-source projects like OpenSSL and BoringSSL have already begun incorporating hybrid key exchange options into their TLS 1.3 implementations, making it possible to test and deploy post-quantum secure communications today.

8.2.5 Performance and Compatibility Considerations

One of the main concerns with integrating PQC into TLS 1.3 is performance. Many post-quantum algorithms have significantly larger key sizes and ciphertexts compared to traditional algorithms. This can increase handshake times and consume more bandwidth, potentially impacting the user experience, especially on mobile or low-bandwidth networks. Additionally, these larger sizes can introduce challenges in constrained environments such as embedded systems or Internet of Things (IoT) devices. Despite these hurdles, ongoing research and optimization efforts have shown that many PQC algorithms are feasible for use in TLS. For instance, Kyber demonstrates competitive performance in both speed and size when compared to traditional methods. Further improvements in implementation efficiency, combined with hardware acceleration, are expected to make PQC integration even more practical in the near future.

8.2.6 Security and Forward Secrecy in a Post-Quantum Context

TLS 1.3 emphasizes the principle of forward secrecy, which ensures that the compromise of long-term keys does not affect the confidentiality of past communications. In the context of quantum threats, this principle becomes even more critical. An adversary could potentially record encrypted traffic today and decrypt it in the future using a quantum computer. By using PQC-based key exchange mechanisms, future sessions can be protected against this type of threat. Hybrid schemes offer a transitional defense, while full adoption of standardized PQC algorithms will provide long-term assurance. TLS 1.3's streamlined architecture and extensibility make it well-suited for incorporating new cryptographic primitives without requiring a complete overhaul of the protocol.

8.2.7 Industry Adoption and Standardization Efforts

Several organizations and companies are already experimenting with TLS 1.3 and PQC. Tech giants like Google and Cloudflare have conducted real-world experiments using hybrid key exchange algorithms in web browsers and servers. These tests help identify practical issues such as latency, error rates, and compatibility problems across diverse platforms. Moreover, the National Security Agency (NSA) in the United States has announced plans to transition to quantum-resistant algorithms as part of its Commercial National Security Algorithm Suite 2.0 (CNSA 2.0). These initiatives underscore the urgency of adopting PQC and the importance of collaboration between academia, industry, and government. Standards bodies such as IETF are also working on defining extensions to TLS that support post-quantum cryptographic mechanisms, ensuring that future versions of the protocol are resilient to quantum attacks.

8.3 INTERNET OF THINGS (IOT) AND EMBEDDED SYSTEMS

The implementation of post-quantum cryptography in IoT and embedded systems represents a critical component of the transition to a quantum-safe world. As organizations prepare for the quantum computing era, comprehensive strategies for securing constrained environments provide the necessary framework for maintaining digital security despite resource limitations. The complex nature of these implementations requires careful optimization and innovative approaches to ensure that security improvements do not come at the expense of operational functionality.

The future of quantum-safe IoT and embedded system security lies in the balanced development of cryptographic solutions that address both security requirements and performance constraints. As quantum computing capabilities continue to advance, the cryptographic community's focus on optimized implementations for constrained environments will remain crucial for enabling practical deployment of quantum-resistant security solutions.

The experiences gained through current implementation efforts will inform future developments, ensuring that IoT and embedded systems can effectively adapt to meet the challenges of evolving computational capabilities while maintaining robust protection in an era of rapidly advancing technological threats. The principles and practices of post-quantum cryptography in constrained environments provide the foundation for securing the digital future of resource-limited devices. The future development of post-quantum cryptography for IoT and embedded systems involves several promising directions that will shape the evolution of quantum-resistant security solutions. The integration of hardware-software co-design approaches presents opportunities for developing cryptographic implementations that are optimized for specific device capabilities.

The development of adaptive cryptographic algorithms that can adjust their security parameters based on available resources or specific threat environments may provide enhanced protection while maintaining performance characteristics. The evolution of standardized interfaces for cryptographic operations could facilitate broader adoption and interoperability between different constrained devices.

The future directions extend to various aspects of IoT and embedded system security. The development of energy-efficient cryptographic implementations that minimize power consumption while maintaining security guarantees becomes particularly important for battery-powered devices. The integration of machine learning techniques for optimizing cryptographic parameters based on specific device characteristics and operational constraints may provide enhanced performance while maintaining robust protection.

The evolution of lightweight cryptographic protocols that are specifically designed for constrained environments will help ensure that post-quantum security can be effectively implemented in resource-limited devices. These future directions will shape the development of quantum-safe IoT and embedded system security, ensuring that these constrained environments can adapt to meet the challenges of advancing computational capabilities while maintaining robust protection in an evolving technological landscape.

8.3.1 Challenges of Post-Quantum Cryptography in Constrained Environments

The implementation of post-quantum cryptography in IoT and embedded systems faces several significant challenges that distinguish these environments from traditional computing platforms. The computational requirements of many quantum-resistant algorithms often exceed the processing capabilities of resource-constrained devices, potentially introducing performance bottlenecks that could disrupt critical operations.

The memory demands of post-quantum algorithms, with their larger key sizes and expanded parameter requirements, can strain the limited storage capacities of embedded systems. These constraints necessitate careful optimization of cryptographic implementations to ensure they can operate effectively within the strict limitations of IoT and embedded environments. Another significant challenge involves the energy constraints of many IoT and embedded devices, where cryptographic operations must minimize power consumption to preserve battery life and maintain operational efficiency. The implementation of post-quantum algorithms must consider these energy limitations while still providing robust security guarantees. The diversity of IoT and embedded platforms introduces additional complexity, as different devices may have varying capabilities and constraints that require tailored cryptographic solutions. These challenges require innovative approaches to post-quantum cryptography that can accommodate the unique requirements of constrained environments while maintaining strong security protections against both classical and quantum threats.

8.3.2 Optimized Cryptographic Algorithms for IoT

The development of optimized cryptographic algorithms forms a fundamental approach to implementing post-quantum security in IoT and embedded systems. These optimized algorithms typically involve simplified mathematical operations that reduce computational requirements while maintaining security guarantees. The optimization process often focuses on the specific characteristics of post-quantum algorithms that can be adapted to operate more efficiently in constrained environments.

For instance, lattice-based cryptographic schemes can be optimized through careful parameter selection that balances security requirements with performance limitations. The development of lightweight variants of post-quantum algorithms that maintain security while reducing computational overhead represents another important optimization approach.

The optimization of cryptographic algorithms extends to various aspects of their implementation in IoT and embedded systems. The selection of algorithm parameters that provide adequate security while minimizing computational requirements forms a critical consideration. The development of efficient arithmetic operations that can be executed with limited processing resources helps improve performance in constrained environments. The optimization process must also consider the memory access patterns of cryptographic operations, where efficient data structures and access methods can significantly impact performance. These comprehensive optimization approaches help ensure that post-quantum cryptographic algorithms can be effectively implemented in resource-constrained environments while maintaining robust security guarantees.

8.3.3 Hardware Acceleration Techniques for Embedded Systems

Hardware acceleration techniques provide essential capabilities for implementing post-quantum cryptography in embedded systems by offloading computationally intensive operations to specialized hardware. These techniques typically involve the development of custom cryptographic co-processors that can efficiently execute the complex mathematical operations characteristic of quantum-resistant algorithms. The implementation of hardware acceleration can significantly improve the performance of post-quantum cryptographic operations while reducing the computational burden on main processors. This approach becomes particularly valuable in embedded systems where processing resources are severely constrained, allowing for the execution of complex cryptographic functions that would otherwise be impractical.

The development of hardware acceleration techniques involves several important considerations specific to embedded environments. The design of specialized arithmetic units that can efficiently execute the particular operations required by post-quantum algorithms helps improve performance while minimizing power consumption. The implementation of memory architectures optimized for cryptographic operations can reduce the overhead associated with large key sizes and expanded parameter requirements. The integration of hardware security features that protect against physical attacks becomes particularly important in embedded systems where devices may be more vulnerable to tampering. These hardware acceleration techniques help ensure that post-quantum cryptographic implementations can operate effectively in resource-constrained environments while maintaining strong security guarantees.

8.3.4 Protocol Optimization for IoT Communications

The optimization of communication protocols forms a critical aspect of implementing post-quantum cryptography in IoT environments, where bandwidth and energy constraints require efficient cryptographic operations. These optimized protocols typically involve the development of lightweight cryptographic schemes that minimize the communication overhead associated with quantum-resistant algorithms. The protocol design process often focuses on reducing the number of messages required for secure communications while maintaining security guarantees. The development of efficient key exchange mechanisms that minimize the computational and communication requirements of establishing secure sessions helps improve performance in constrained environments.

The protocol optimization extends to various aspects of IoT communications. The design of compact message formats that reduce the bandwidth requirements of cryptographic operations helps accommodate the limited communication capabilities of many IoT devices. The development of efficient authentication mechanisms that provide strong security guarantees while minimizing computational overhead becomes particularly important in resource-constrained environments. The optimization of protocol parameters that balance security requirements with performance limitations helps ensure that post-quantum cryptographic implementations can operate effectively in IoT communications while maintaining robust protection against both classical and quantum threats.

8.3.5 Key Management Strategies for Constrained Devices

The management of cryptographic keys in IoT and embedded systems requires specialized strategies that accommodate the unique constraints of these environments. These key management approaches typically involve the development of efficient key generation and storage mechanisms that minimize memory requirements while maintaining security guarantees. The implementation of hierarchical key management systems that can derive multiple keys from a single master key helps reduce storage demands in constrained devices. The development of key update mechanisms that can refresh cryptographic material without requiring complete rekeying operations helps maintain security while minimizing computational overhead.

The key management strategies extend to various aspects of IoT and embedded system operations. The implementation of secure key storage mechanisms that protect cryptographic material from physical attacks becomes particularly important in devices that may be vulnerable to tampering. The development of efficient key distribution protocols that minimize the communication overhead associated with key exchange helps improve performance in constrained environments. The design of key lifecycle management systems that can handle the larger key sizes characteristic of post-quantum algorithms while operating within memory limitations helps ensure that cryptographic implementations can maintain security in resource-constrained devices.

8.3.6 Side-Channel Resistance in Embedded Implementations

The implementation of side-channel resistant techniques forms a critical aspect of secure post-quantum cryptography in embedded systems, where physical attacks can exploit implementation characteristics to compromise security. These resistance techniques typically involve the development of constant-time algorithms that execute cryptographic operations in uniform time regardless of secret values. The implementation of masking techniques that obscure intermediate values during computation helps prevent information leakage through power analysis or electromagnetic emissions. The design of secure memory access patterns that minimize data-dependent timing variations helps protect against cache attacks that might exploit memory operations to infer sensitive information.

The side-channel resistance techniques extend to various aspects of embedded system implementations. The development of hardware security features that protect against physical tampering becomes particularly important in devices that may be vulnerable to direct attacks. The implementation of secure boot mechanisms that verify the integrity of cryptographic implementations helps prevent modification of security functions. The design of power management techniques that minimize variations in energy consumption during cryptographic operations helps protect against power analysis attacks. These comprehensive side-channel resistance approaches help ensure that post-quantum cryptographic implementations in embedded systems maintain robust protection against both mathematical and physical attacks.

8.3.7 Testing and Validation Methodologies

Comprehensive testing and validation form essential components of secure post-quantum cryptographic implementations in IoT and embedded systems. The testing process typically begins with functional verification that confirms cryptographic operations behave as expected under the resource constraints of constrained environments.

This involves evaluating both the mathematical correctness of post-quantum implementations and their operational characteristics in specific hardware configurations. Security validation forms a critical component of the testing process, assessing resistance to both mathematical and implementation attacks across different algorithm configurations. The validation process must consider the unique characteristics of IoT and embedded environments, where different devices may present specific security requirements and potential vulnerabilities.

The testing methodologies extend to various aspects of constrained system operations. Performance profiling evaluates the computational and memory impacts of different algorithm configurations to ensure that security updates do not introduce unacceptable overhead. Power consumption analysis measures the energy requirements of cryptographic operations to verify they remain within the limitations of battery-powered devices. Side-channel analysis examines implementations for physical vulnerabilities that might leak sensitive information through execution characteristics. These comprehensive testing approaches help ensure that post-quantum cryptographic implementations in IoT and embedded systems maintain robust protection while functioning correctly in operational environments.

8.4 BLOCKCHAIN AND CRYPTOCURRENCY RESILIENCE

The blockchain industry stands at a critical juncture where proactive measures can preempt quantum threats. By integrating post-quantum cryptography through hybrid models, collaborative research, and decentralized governance, networks can preserve their security and trustless nature in the face of advancing quantum computing. While challenges such as scalability and backward compatibility remain, the growing focus on PQC standardization and implementation signals a collective commitment to resilience. For investors, developers, and users, the message is clear: the time to future-proof blockchain is now. Embracing post-quantum cryptography is not merely a technical upgrade but a necessary evolution to ensure that decentralized systems remain robust, secure, and viable in the decades to come.

8.4.1 Quantum Vulnerabilities in Blockchain

The primary vulnerabilities in blockchain networks stem from two quantum algorithms: Shor's algorithm and Grover's algorithm. Shor's algorithm can efficiently solve the integer factorization and discrete logarithm problems, which underpin RSA and ECC, respectively. This capability means that an attacker with a sufficiently powerful quantum computer could derive private keys from public keys, compromising wallet security and transaction integrity. Grover's algorithm, while less destructive, offers a quadratic speedup in brute-force searches, effectively halving the security strength of symmetric cryptographic systems like SHA-256. For blockchain networks, this implies that quantum computers could theoretically reverse-engineer hashed data or weaken mining difficulty adjustments. The combination of these threats necessitates a shift toward quantum-resistant cryptographic primitives in blockchain design.

8.4.2 Post-Quantum Cryptography Solutions for Blockchain

To mitigate quantum risks, researchers and developers are exploring post-quantum cryptographic algorithms that remain secure against quantum attacks. Lattice-based cryptography, hash-based signatures, and multivariate polynomial cryptography are among the leading candidates for PQC standardization by the National Institute of Standards and Technology (NIST). Lattice-based schemes, such as CRYSTALS-Kyber and CRYSTALS-Dilithium, offer strong security with relatively efficient key sizes, making them suitable for blockchain applications. Hash-based signatures, like SPHINCS+, provide quantum resistance but require larger signature sizes, which could impact transaction throughput. Multivariate and code-based cryptography are also being considered, though they present trade-offs in computational complexity and adoption feasibility. Integrating these algorithms into blockchain protocols will require careful evaluation to balance security, performance, and decentralization.

8.4.3 Challenges in Implementing Post-Quantum Blockchain

Transitioning blockchain systems to post-quantum cryptography presents several technical and logistical challenges. One major obstacle is the increased computational and storage overhead associated with PQC algorithms. Many quantum-resistant schemes demand larger key sizes and signature lengths, which could inflate transaction data and strain network scalability. For example, Bitcoin's block size limitations may conflict with the expanded data requirements of post-quantum signatures. Another challenge is backward compatibility—existing wallets and smart contracts rely on classical cryptography, and a sudden shift could disrupt functionality and user access. Additionally, achieving consensus among decentralized networks for cryptographic upgrades is inherently complex, as it requires coordinated hard forks and broad stakeholder agreement. These hurdles necessitate a phased, hybrid approach to PQC adoption in blockchain ecosystems.

8.4.4 Hybrid Cryptographic Approaches for a Smoother Transition

To address compatibility and performance concerns, many blockchain projects are considering hybrid cryptographic models that combine classical and post-quantum algorithms. In this approach, transactions could be secured using both ECC and a PQC scheme, ensuring that even if one system is compromised, the other remains intact. Ethereum and other smart contract platforms are exploring modular cryptographic designs that allow for gradual PQC integration without breaking existing applications. Hybrid solutions also mitigate risks during the transition period, as quantum computers are not yet capable of breaking classical cryptography at scale. By incrementally introducing PQC components—such as quantum-resistant signature schemes or key exchange mechanisms—blockchain networks can enhance resilience while maintaining operational continuity.

8.4.5 Industry Initiatives and Quantum-Resistant Blockchains

Several blockchain projects and research initiatives are already pioneering quantum-resistant solutions. The Quantum Resistant Ledger (QRL) is a blockchain designed exclusively with PQC, utilizing hash-based XMSS signatures for transaction security. Algorand has proposed state-proof mechanisms incorporating post-quantum signatures to future-proof its consensus protocol. Meanwhile, Ethereum's research team is evaluating NIST-approved PQC algorithms for potential integration into its roadmap. Beyond individual projects, consortiums like the Post-Quantum Cryptography Alliance (PQCA) are fostering collaboration between academia and industry to accelerate PQC standardization for decentralized systems. These efforts highlight the blockchain community's recognition of quantum threats and its commitment to proactive mitigation.

8.4.6 Role of Decentralized Governance in Cryptographic Upgrades

Unlike traditional systems, blockchain networks rely on decentralized governance to implement protocol changes. Upgrading cryptographic standards in a decentralized environment requires broad consensus among developers, miners, node operators, and users. Hard forks—such as those used to introduce SegWit in Bitcoin or Ethereum's transition to proof-of-stake—demonstrate the complexities of coordinating large-scale updates. For PQC adoption, blockchain communities must navigate technical debates, stakeholder incentives, and potential chain splits. Transparent communication, robust testing (e.g., through testnets), and incentive alignment will be critical to ensuring smooth transitions. Decentralized autonomous organizations (DAOs) and on-chain governance mechanisms may play a pivotal role in facilitating these upgrades while preserving network integrity.

8.4.7 Long-Term Implications for Cryptocurrencies and Decentralized Finance

The successful integration of post-quantum cryptography will determine the long-term viability of cryptocurrencies and decentralized finance (DeFi) in a quantum-powered world. Without PQC, quantum attacks could undermine trust in blockchain-based assets, leading to market instability or regulatory intervention. Conversely, early adopters of quantum-resistant blockchains may gain a competitive advantage by positioning themselves as secure and future-proof. DeFi protocols, which rely heavily on smart contract security, must also prioritize PQC to protect user funds and prevent exploits. As central banks explore digital currencies (CBDCs), quantum resilience will likely become a regulatory requirement, further incentivizing the industry to adopt PQC standards.

8.5 GOVERNMENT, MILITARY, AND CRITICAL INFRASTRUCTURE SYSTEMS

The transition to post-quantum cryptography (PQC) represents a defining moment in the realm of digital security, particularly for government, military, and critical infrastructure systems. These sectors form the backbone of national security and public welfare, relying heavily on robust cryptographic mechanisms to ensure the confidentiality, integrity, and availability of their information and communication systems.

As quantum computing technology evolves and threatens the classical cryptographic algorithms in use today, there is an urgent need to adopt cryptographic solutions resilient to quantum attacks. The strategic integration of post-quantum cryptography into these domains is not merely a technical upgrade but a critical component of long-term security planning.

Government institutions are among the largest generators and protectors of sensitive data. From classified diplomatic communications to citizen identity management and tax records, they maintain vast digital repositories that must be secured against both present and future threats. The current cryptographic infrastructure is built on public-key cryptography schemes such as RSA, DSA, and ECC. These schemes are highly vulnerable to Shor's algorithm, a quantum computing algorithm capable of solving the mathematical problems underpinning traditional cryptography. A sufficiently advanced quantum computer could break these algorithms, thereby exposing state secrets, compromising national identities, and disrupting core services. Post-quantum cryptographic algorithms, designed to resist both classical and quantum attacks, are essential to maintaining digital sovereignty and public trust.

In the military domain, the consequences of a cryptographic breach could be catastrophic. Military communication systems, command and control networks, satellite communication, intelligence operations, and autonomous weapon systems all rely on secure and authenticated data exchange. The confidentiality of operational planning and the integrity of strategic communications are vital to mission success and national defense. Adversaries with access to quantum computing capabilities could decrypt intercepted communications, spoof identities, or manipulate battlefield data. The military's adoption of post-quantum cryptographic standards is therefore crucial to maintaining a competitive edge and deterring future threats. Implementation must consider both legacy systems and new platforms, ensuring that post-quantum security is embedded across all levels of defense architecture.

Critical infrastructure systems include energy grids, transportation networks, water supply systems, financial networks, healthcare infrastructure, and telecommunication systems. These systems depend on real-time data exchange and control, often facilitated through supervisory control and data acquisition (SCADA) systems and industrial control systems (ICS). Any disruption in these networks can lead to cascading failures, causing massive economic losses and endangering lives.

The security of these systems has traditionally relied on public-key infrastructure (PKI) for authentication, secure communication, and software updates. With the arrival of quantum computing, the digital certificates and encryption protocols used in these infrastructures become vulnerable to compromise. Post-quantum cryptography ensures that even if adversaries gain access to quantum computers, they cannot easily decrypt communications or forge control commands within critical infrastructure systems.

A major challenge in transitioning to PQC in government, military, and critical infrastructure systems lies in the complexity and diversity of these environments. These sectors operate with long lifecycles and strict regulatory frameworks. Upgrading cryptographic systems involves retrofitting legacy systems, retraining personnel, and conducting comprehensive testing to ensure reliability and performance. For instance, government agencies using smart ID cards or secure voting systems must ensure that the underlying cryptographic protocols can be updated without compromising backward compatibility. Similarly, the military must test PQC implementations in mission-critical scenarios to evaluate latency, power consumption, and resistance to side-channel attacks. Infrastructure operators must assess the impact of larger key sizes and increased computational demands on system efficiency and responsiveness.

International standardization efforts play a central role in enabling secure transitions to post-quantum cryptography. The National Institute of Standards and Technology (NIST) has taken the lead in standardizing quantum-resistant cryptographic algorithms through its Post-Quantum Cryptography Standardization project. The ongoing selection and evaluation process involves extensive testing, public scrutiny, and peer review. Several promising candidates have emerged, including Kyber for key encapsulation and Dilithium for digital signatures. These algorithms are being considered for integration into protocols such as TLS, SSH, and IPsec, which are widely used across government, defense, and infrastructure sectors. Compliance with these standards ensures interoperability, reduces the risk of implementation flaws, and facilitates international cooperation in secure communications.

Governments are increasingly developing and releasing guidelines to facilitate post-quantum readiness. For example, the U.S. National Security Agency (NSA) has published the Commercial National Security Algorithm Suite 2.0 (CNSA 2.0), which outlines a roadmap for transitioning to quantum-resistant algorithms in national security systems. This includes recommendations for secure key exchanges, digital signatures, and software signing processes. Other countries, such as Germany and the United Kingdom, are also investing in quantum-safe strategies, including research funding, policy development, and public-private partnerships. These initiatives reflect a growing consensus on the need to prepare national digital assets for a quantum future.

From a technological perspective, the integration of post-quantum cryptography requires robust hardware and software support. Cryptographic libraries must be updated to support new algorithms while ensuring performance and resilience. Hardware security modules (HSMs), smart cards, and trusted platform modules (TPMs) must be evaluated for compatibility with post-quantum operations. Additionally, cloud-based services and edge computing platforms used in critical infrastructure must be configured to use quantum-safe cryptographic mechanisms. Secure update mechanisms, especially in embedded and IoT devices, are crucial to enable remote patching of cryptographic components in response to emerging threats or vulnerabilities.

CHAPTER 9

Policy, Compliance, and Global Perspectives

9.1 REGULATORY LANDSCAPE AND NATIONAL POLICIES

The development of regulatory frameworks and national policies for post-quantum cryptography represents a critical component of the transition to quantum-resistant security. As governments and regulatory bodies establish standards for cryptographic implementations, these policies provide the necessary guidance for organizations to adopt quantum-safe solutions while maintaining compliance with security requirements. The complex nature of this regulatory landscape requires careful coordination between technical standards, national security considerations, and international cooperation efforts.

The future of cryptographic regulation lies in the balanced development of policies that address both security requirements and practical implementation considerations. As quantum computing capabilities continue to advance, the regulatory community's focus on establishing comprehensive frameworks for post-quantum cryptography will remain crucial for enabling secure global communications.

The experiences gained through current regulatory efforts will inform future policy developments, ensuring that cryptographic regulations can adapt to meet the challenges of evolving computational capabilities while maintaining robust protection in the quantum era. The establishment of clear regulatory landscapes provides the foundation for building a secure digital future in an era of rapidly advancing technological threats.

The future development of post-quantum cryptography regulations involves several promising directions that will shape the evolution of security policies. The integration of adaptive regulatory frameworks that can respond to emerging quantum threats represents an important future direction. These frameworks would incorporate mechanisms for updating cryptographic requirements as new vulnerabilities are discovered or computational capabilities advance. The development of standardized compliance assessment methodologies could facilitate more consistent evaluations of cryptographic implementations across different jurisdictions. The establishment of international certification mutual recognition agreements may help create more harmonized regulatory environments for global cryptographic implementations. The future directions extend to various aspects of cryptographic policy development. The incorporation of performance metrics into regulatory requirements could help ensure that security implementations meet both protection and operational standards. The development of sector-specific guidelines that address the unique requirements of different industries may provide more tailored regulatory approaches.

The establishment of continuous monitoring requirements for cryptographic implementations could help maintain security as threats evolve. These future directions will shape the development of post-quantum cryptography regulations, ensuring that security policies can adapt to meet the challenges of advancing computational capabilities while maintaining robust protection in an evolving technological landscape.

9.1.1 International Standards and Regulatory Frameworks

The foundation of post-quantum cryptography regulation rests on international standards that provide technical specifications for quantum-resistant algorithms. Organizations like the National Institute of Standards and Technology (NIST) in the United States and the International Organization for Standardization (ISO) play pivotal roles in developing these technical standards that form the basis for regulatory frameworks. The NIST Post-Quantum Cryptography Standardization Project has emerged as a global reference point, with its selected algorithms serving as benchmarks for national policies. These international standards provide the technical underpinnings that enable regulatory bodies to establish consistent requirements for cryptographic implementations across different jurisdictions.

The development of regulatory frameworks builds upon these technical standards to create enforceable policies for post-quantum cryptography adoption. These frameworks typically establish timelines for migration, specify approved algorithms, and define compliance requirements for different industry sectors. The regulatory process involves public consultation periods that allow stakeholders to provide input on proposed policies, helping ensure that resulting regulations are both effective and practical. The harmonization of these frameworks across national boundaries becomes increasingly important as global communications require consistent security protections. This international coordination helps prevent regulatory fragmentation that could create vulnerabilities in cross-border data flows.

9.1.2 National Security Considerations

National security considerations form a fundamental aspect of post-quantum cryptography policies, as governments recognize the strategic importance of maintaining secure communications in the quantum era. The development of these policies involves assessments of potential threats to national infrastructure and sensitive communications that could be compromised by quantum computing advances. Many nations have established dedicated task forces to evaluate the implications of quantum computing for national security and develop appropriate responses. These considerations often lead to prioritized adoption timelines for government systems and critical infrastructure that handle particularly sensitive information.

The national security aspects of cryptographic policy extend to various dimensions of implementation. Export control regulations for cryptographic technologies often require updates to address post-quantum algorithms, with governments establishing new frameworks for managing these security-sensitive technologies. The protection of classified information systems typically involves accelerated migration timelines to ensure continued security against quantum threats.

The development of national cryptographic standards that may differ from international specifications creates additional complexity in the regulatory landscape. These national security considerations help shape post-quantum cryptography policies that address both immediate threats and long-term strategic requirements.

9.1.3 Industry-Specific Regulatory Approaches

The regulatory landscape for post-quantum cryptography involves industry-specific approaches that address the unique requirements and challenges of different sectors. Financial services regulations typically establish early adoption timelines for post-quantum algorithms, recognizing the sensitivity of financial data and transaction systems.

Healthcare industry policies focus on protecting patient information through secure cryptographic implementations while maintaining compliance with privacy regulations. The defense sector often develops specialized cryptographic standards that address military and intelligence communications requirements. These industry-specific approaches help ensure that post-quantum cryptography adoption meets the particular needs of each sector while maintaining overall security standards.

The industry regulations extend to various aspects of cryptographic implementation. Compliance requirements for data protection often involve specific cryptographic standards that organizations must meet to ensure adequate security. The certification processes for cryptographic modules typically require updates to address post-quantum algorithms and their implementations. Audit procedures for security systems frequently incorporate evaluations of cryptographic practices to verify compliance with regulatory requirements. These industry-specific regulatory approaches help create a comprehensive framework for post-quantum cryptography adoption that addresses the diverse needs of different economic sectors while maintaining consistent security standards.

9.1.4 Compliance Timelines and Migration Strategies

The establishment of compliance timelines forms a critical component of post-quantum cryptography regulations, providing organizations with clear guidance on migration requirements. These timelines typically involve phased approaches that prioritize updates to the most vulnerable systems while allowing for gradual adoption across less critical implementations. The regulatory frameworks often specify different deadlines for various sectors based on their security requirements and operational constraints. Early compliance periods frequently include provisions for testing and validation of post-quantum implementations before full enforcement begins. These structured timelines help organizations plan their cryptographic transitions while ensuring that security improvements occur within reasonable periods.

The migration strategies incorporated into regulatory frameworks address various aspects of cryptographic transitions. Compatibility requirements often specify that systems must maintain interoperability during transition periods when both classical and post-quantum algorithms may be in use. The regulatory approaches typically include provisions for cryptographic agility that allow systems to adapt to evolving standards and new algorithm developments. Compliance verification processes help ensure that organizations meet regulatory requirements through audits and certification procedures. These comprehensive migration strategies provide organizations with clear pathways for transitioning to post-quantum cryptography while maintaining operational continuity and regulatory compliance.

9.1.5 Certification and Validation Processes

Certification processes for post-quantum cryptographic implementations form essential components of regulatory frameworks, providing mechanisms to verify that security solutions meet established standards. These certification programs typically involve rigorous testing of cryptographic modules to evaluate their security characteristics and implementation quality.

The validation processes often include cryptographic algorithm verification that confirms implementations correctly follow specified standards. Side-channel resistance testing evaluates implementations for vulnerabilities that might leak sensitive information through physical characteristics. These certification processes help build confidence in post-quantum cryptographic solutions while ensuring they meet regulatory security requirements.

The certification frameworks extend to various aspects of cryptographic implementation. Evaluation criteria for cryptographic modules typically specify security requirements that implementations must satisfy to receive certification. The testing methodologies often incorporate both functional verification that confirms correct operation and security validation that assesses resistance to various attack vectors. Certification laboratories accredited to perform these evaluations provide independent verification of cryptographic implementations. These comprehensive certification processes help ensure that post-quantum cryptographic solutions meet regulatory standards while providing organizations with verified security implementations.

9.1.6 International Cooperation and Policy Harmonization

International cooperation plays a crucial role in developing consistent regulatory approaches to post-quantum cryptography, as global communications require harmonized security standards. Multilateral agreements between nations help establish common frameworks for cryptographic regulations that facilitate secure cross-border data flows. International standards organizations provide forums for coordinating policy development and sharing best practices among regulatory bodies.

These cooperative efforts aim to prevent regulatory fragmentation that could create vulnerabilities in global communications networks. The harmonization of cryptographic policies becomes particularly important for multinational organizations that must comply with different national regulations while maintaining consistent security protections.

The international cooperation extends to various dimensions of cryptographic regulation. Mutual recognition agreements for cryptographic certifications help ensure that implementations verified in one jurisdiction are accepted in others. The development of common evaluation criteria for cryptographic modules facilitates consistent security assessments across different regulatory regimes. Information sharing arrangements between national security agencies help coordinate responses to emerging quantum threats. These international cooperation efforts help create a more consistent regulatory landscape for post-quantum cryptography that supports secure global communications while accommodating different national requirements.

9.2 DATA PROTECTION LAWS AND PQC

The emergence of post-quantum cryptography (PQC) has prompted a significant re-evaluation of global data protection practices in light of future quantum threats. Data protection laws, which are designed to safeguard the privacy and security of personal and sensitive data, must now consider not only current threats but also the potential capabilities of quantum computers. Quantum computers, once they reach a sufficient level of maturity, could render current cryptographic algorithms obsolete, particularly those that underpin public-key infrastructures. As a result, integrating PQC into legal, regulatory, and technological frameworks has become a pressing issue for governments, organizations, and privacy advocates around the world.

Data protection laws serve as the legal backbone for ensuring that individuals' personal data is collected, processed, and stored securely and ethically. Regulations such as the European Union's General Data Protection Regulation (GDPR), the California Consumer Privacy Act (CCPA), and other similar frameworks worldwide emphasize the principles of data minimization, lawful processing, consent, transparency, and most importantly, security of data. These laws mandate that organizations implement appropriate technical and organizational measures to protect personal data from unauthorized access, loss, or destruction. Cryptographic security plays a central role in fulfilling these obligations, with encryption being explicitly recognized in many of these regulations as a preferred method of protecting data.

Traditional encryption methods, such as RSA and elliptic-curve cryptography (ECC), are currently used to secure data at rest, data in transit, and data in use. These algorithms provide the foundational elements for digital certificates, secure key exchanges, digital signatures, and encrypted communication. However, these classical algorithms are based on mathematical problems that quantum computers can efficiently solve using algorithms like Shor's and Grover's. This renders the current public-key encryption ecosystem vulnerable to future quantum attacks. The threat is not just theoretical. The notion of "harvest now, decrypt later" highlights the risk that encrypted data being transmitted or stored today could be collected by adversaries and decrypted in the future once quantum capabilities become practical.

In the context of data protection laws, this looming threat raises critical concerns about compliance and risk management. For example, the GDPR requires that personal data be protected in a way that ensures security appropriate to the risk, including the pseudonymization and encryption of data. The regulation also mandates that organizations perform Data Protection Impact Assessments (DPIAs) when introducing new technologies that could impact privacy. Failing to anticipate the quantum threat and not taking steps toward PQC integration could be seen as a failure to implement adequate security measures, potentially resulting in legal consequences, fines, and loss of trust.

Post-quantum cryptography refers to cryptographic algorithms that are secure against both classical and quantum attacks. These algorithms are being standardized by the National Institute of Standards and Technology (NIST) in the United States, with international participation and scrutiny. Among the most prominent candidates selected for standardization are Kyber for key encapsulation and CRYSTALS-Dilithium for digital signatures. These algorithms are based on hard mathematical problems such as lattice-based, multivariate, code-based, and hash-based constructions, which are believed to be resistant to known quantum attacks. Integrating these algorithms into existing infrastructures is essential for ensuring long-term data protection and legal compliance.

For lawmakers and regulators, the transition to PQC introduces both challenges and opportunities. One key challenge is the pace of change. Data protection laws often lag behind technological advancements, and the process of updating legislation can be slow. However, some regulatory bodies have begun acknowledging the need to prepare for quantum computing. For instance, the European Union Agency for Cybersecurity (ENISA) and the U.S. National Security Agency (NSA) have both issued guidance and roadmaps for quantum-resistant security practices. There is also a growing international dialogue around quantum readiness, urging regulatory authorities to update security frameworks, compliance checklists, and certification schemes to incorporate PQC standards.

Another legal dimension relates to data sovereignty and cross-border data transfers. Data protection laws often include strict provisions about the geographical location and jurisdiction under which personal data is processed. As PQC algorithms and implementations become part of national security policies and strategic technology roadmaps, the control over cryptographic tools may become more tightly regulated. Governments may impose restrictions on the export or import of cryptographic technology, including PQC tools, especially if they are integrated into critical sectors. This may affect international cloud providers, multinational corporations, and global service providers that rely on cross-border data exchange and global cryptographic infrastructure.

From a corporate compliance standpoint, integrating PQC into enterprise systems involves substantial operational planning. Organizations must evaluate their existing cryptographic assets, assess their vulnerabilities to quantum attacks, and develop migration strategies. This includes updating secure communication protocols such as TLS, VPN, and SSH, redesigning PKI systems, and ensuring that digital signatures used for legal documents, contracts, and identity verification are quantum-resistant. These changes must be documented and justified within legal compliance audits, demonstrating due diligence in protecting data in accordance with regulatory expectations.

Another aspect of the legal implications of PQC is data retention and archiving. Many data protection laws mandate that personal data not be kept longer than necessary and must be securely deleted or anonymized when no longer in use. However, in sectors like healthcare, finance, and government, records may be retained for decades or even indefinitely. If these archives are encrypted with classical algorithms, they could be vulnerable to decryption by future quantum computers. Ensuring that long-term data storage complies with quantum-resilient encryption standards is therefore vital for legal and ethical data stewardship.

Legal frameworks must also address the certification and validation of PQC algorithms. Just as current cryptographic modules are subject to certification under programs like the Federal Information Processing Standards (FIPS) in the U.S., post-quantum cryptographic solutions must be vetted, standardized, and certified to ensure that they meet legal, technical, and interoperability requirements. Regulators and certification bodies will need to define the criteria for PQC adoption, including acceptable key lengths, algorithm parameters, and usage scenarios across different sectors.

Privacy-enhancing technologies (PETs), such as homomorphic encryption and zero-knowledge proofs, are also being reconsidered through the lens of PQC. These technologies are integral to privacy-focused legal regimes and are often used in compliance scenarios involving minimal data exposure, such as secure multi-party computation and anonymized analytics. Ensuring that PETs are built upon quantum-secure foundations is essential for preserving their legal and practical utility in a post-quantum era.

Education and legal literacy among stakeholders is crucial for the success of PQC adoption. Lawmakers, judges, regulators, and corporate compliance officers must understand the basics of quantum computing and its implications for data protection. Legal education and continuing professional development programs must include content on PQC and emerging cryptographic standards to ensure informed decision-making in legislative drafting, contract formulation, and regulatory enforcement.

9.3 INTEROPERABILITY AND CROSS-BORDER COORDINATION

The transition to post-quantum cryptography is not just a technical upgrade but a global imperative that demands unprecedented levels of cooperation. Interoperability and cross-border coordination are essential to preventing cryptographic fragmentation and maintaining secure digital ecosystems in the face of quantum threats.

By aligning standardization efforts, fostering international research collaboration, and developing harmonized regulatory frameworks, stakeholders can ensure a smooth and inclusive transition to quantum-resistant security. The success of this endeavor will depend on the willingness of nations, industries, and institutions to prioritize collective security over isolated advancements. In doing so, the global community can build a resilient foundation for the future of cryptography—one that upholds privacy, trust, and connectivity across borders in the quantum era.

The path to achieving interoperability in post-quantum cryptography will require sustained effort across multiple fronts. Continued research into optimizing PQC algorithms for performance and scalability will make them more viable for widespread adoption. International working groups should establish testing and certification processes to validate PQC implementations across different use cases. Public-private partnerships can accelerate the deployment of quantum-resistant infrastructures, particularly in regions with limited resources. Additionally, cybersecurity education and awareness campaigns will be necessary to prepare organizations and policymakers for the transition. As quantum computing evolves, so too must the frameworks for global cryptographic coordination, ensuring that security, trust, and interoperability remain intact in a post-quantum world.

9.3.1 Role of Standardization Bodies in PQC Interoperability

Standardization organizations such as the National Institute of Standards and Technology (NIST), the International Organization for Standardization (ISO), and the European Telecommunications Standards Institute (ETSI) play a pivotal role in shaping the future of post-quantum cryptography. NIST's ongoing PQC standardization project has already identified promising algorithms like CRYSTALS-Kyber for encryption and CRYSTALS-Dilithium for digital signatures, which are expected to form the foundation of quantum-resistant security.

However, for these standards to achieve global interoperability, they must be adopted and implemented consistently across different regions and industries. Divergent approaches—such as countries developing their own proprietary PQC solutions—could lead to compatibility issues, complicating international communications and commerce. A unified framework, supported by multilateral agreements, is essential to prevent cryptographic fragmentation and ensure seamless cross-border operations.

9.3.2 Challenges in Achieving Cryptographic Interoperability

One of the primary obstacles to PQC interoperability is the varying levels of technological readiness and regulatory priorities among nations. Some countries may accelerate PQC adoption due to heightened security concerns, while others might lag due to resource constraints or competing policy agendas. Additionally, legacy systems that rely on classical cryptography will need to coexist with quantum-resistant upgrades during the transition period, creating hybrid environments that must remain interoperable. The increased computational overhead of PQC algorithms also poses challenges for systems with limited bandwidth or processing power, particularly in developing regions. Furthermore, geopolitical tensions could influence cryptographic standards, with some nations favoring domestically developed algorithms over internationally recognized ones. Addressing these challenges requires diplomatic engagement, technical cooperation, and flexible migration strategies that accommodate diverse infrastructures.

9.3.3 Cross-Border Collaboration in Research and Development

The development of post-quantum cryptographic solutions is a global endeavor, with researchers from academia, industry, and government contributing to advancements in the field. Collaborative initiatives, such as the Open Quantum Safe project, provide open-source implementations of PQC algorithms, enabling widespread testing and refinement. International partnerships, like those between the U.S. and EU on quantum technology research, facilitate knowledge sharing and accelerate progress toward standardized solutions. Multinational corporations operating in sectors such as finance, telecommunications, and cloud computing also play a crucial role in driving PQC adoption by integrating quantum-resistant protocols into their global infrastructures. By fostering open collaboration and avoiding siloed approaches, stakeholders can ensure that PQC solutions are both interoperable and resilient against emerging threats.

9.3.4 Regulatory and Policy Frameworks for PQC Harmonization

Governments must establish regulatory frameworks that promote the consistent adoption of post-quantum cryptography while addressing national security and economic interests. Policies that mandate PQC compliance for critical infrastructure, such as banking, healthcare, and energy grids, can create a baseline for interoperability. However, these regulations must be aligned with international norms to avoid conflicting requirements that hinder cross-border operations. For instance, the EU's General Data Protection Regulation (GDPR) and the U.S. National Security Agency's (NSA) PQC guidelines should ideally converge to facilitate data protection and cybersecurity across jurisdictions. Diplomatic forums, such as the United Nations and the G7, can serve as platforms for negotiating cryptographic standards and fostering mutual trust among nations. A coordinated policy approach will be essential to balancing sovereignty, security, and global connectivity in the quantum age.

9.3.5 Industry-Specific Implications of PQC Interoperability

Different sectors face unique challenges in implementing interoperable post-quantum cryptography. In finance, cross-border payment systems like SWIFT and blockchain networks must ensure that quantum-resistant encryption does not disrupt transaction speeds or compliance protocols. Telecommunications providers, which underpin global internet connectivity, need to upgrade cryptographic protocols in ways that maintain compatibility across networks. Cloud service providers, storing and processing vast amounts of international data, must adopt PQC without causing service interruptions for multinational clients. Each industry will require tailored migration strategies, but all must prioritize interoperability to avoid creating isolated security ecosystems. Industry consortia, such as the Cloud Security Alliance and the Quantum Economic Development Consortium (QED-C), can help bridge gaps by developing best practices and certification programs for PQC implementation.

9.3.6 Role of Hybrid Cryptography in Transitional Interoperability

During the shift from classical to post-quantum cryptography, hybrid cryptographic systems—combining traditional and quantum-resistant algorithms—will be vital in maintaining interoperability. Hybrid solutions allow systems to communicate securely even if some components have not yet been upgraded to PQC. For example, a hybrid TLS handshake might use both RSA and a post-quantum key exchange mechanism, ensuring backward compatibility while introducing quantum resistance.

This approach minimizes disruption and provides a safety net against cryptographic failures during the transition. However, hybrid systems also introduce complexity, requiring careful management of key sizes, performance trade-offs, and protocol negotiations. Standardizing hybrid implementations will be crucial to ensuring that they function consistently across borders and industries.

9.4 INCENTIVES AND FUNDING FOR PQC TRANSITION

The establishment of effective incentives and funding mechanisms represents a critical component of the transition to post-quantum cryptography. As organizations and governments prepare for the quantum computing era, comprehensive financial support structures provide the necessary resources for adopting quantum-resistant security solutions. The complex nature of this transition requires careful coordination of funding sources to ensure that security improvements receive adequate financial support while maintaining operational continuity.

The future of cryptographic security lies in the balanced development of funding approaches that address both immediate needs and long-term requirements. As quantum computing capabilities continue to advance, the establishment of effective incentive structures will remain crucial for enabling practical adoption of quantum-resistant security solutions. The experiences gained through current funding efforts will inform future developments, ensuring that financial resources can effectively support the transition to post-quantum cryptography while maintaining robust protection in an era of rapidly advancing technological threats.

The establishment of comprehensive funding mechanisms provides the foundation for building a secure digital future in the quantum era. The future development of funding mechanisms for post-quantum cryptography transitions involves several promising directions that will shape the evolution of security investments. The establishment of public-private partnership models presents opportunities for more efficient allocation of transition resources.

These partnerships could leverage government funding with private sector investments to create more comprehensive support structures. The development of outcome-based funding approaches that tie financial support to measurable security improvements may provide more effective incentives for cryptographic transitions. The future funding directions extend to various aspects of the cryptographic transition. The creation of revolving loan funds could help organizations finance their migrations while repaying from operational savings. The establishment of insurance mechanisms that reduce risk premiums for organizations demonstrating quantum-resistant security capabilities may provide additional incentives.

The development of international funding pools that support global cryptographic transitions could help create more consistent security standards. These future directions will shape the evolution of post-quantum cryptography funding, ensuring that financial resources can effectively support the transition to quantum-resistant security solutions. The establishment of effective funding mechanisms for post-quantum cryptography transitions faces several significant challenges that require careful consideration. The substantial costs associated with cryptographic migrations create funding requirements that may exceed available resources in some sectors. The complexity of modern digital ecosystems complicates the allocation of financial support, as different systems may have varying transition requirements. The need to balance immediate security needs with long-term investments presents another challenge, as funding priorities must address both current vulnerabilities and future threats.

The funding challenges extend to various aspects of the cryptographic transition. The allocation of limited resources requires careful prioritization of different migration needs. The coordination of multiple funding sources creates administrative complexities that must be carefully managed. The evaluation of funding effectiveness presents difficulties in measuring the security benefits of transition investments. These challenges necessitate comprehensive approaches that address both financial constraints and security requirements to ensure effective support for post-quantum cryptography adoption.

9.5 ROLE OF INTERNATIONAL COOPERATION IN CYBERSECURITY

In the digital age, cybersecurity has become one of the most pressing global concerns. With the proliferation of digital infrastructure, cloud computing, internet-of-things (IoT) devices, and global connectivity, securing data and systems against cyber threats is a shared responsibility. The emerging challenge posed by quantum computing adds a new and complex dimension to this landscape. The potential of quantum computers to break widely-used public-key cryptographic systems poses a serious threat to the security and confidentiality of digital communications worldwide. This has given rise to the field of post-quantum cryptography (PQC), which focuses on developing cryptographic algorithms that are resistant to quantum attacks.

Quantum computing has introduced the possibility of a paradigm shift in information security. Current cryptographic systems, such as RSA and elliptic curve cryptography, rely on the computational difficulty of certain mathematical problems. These problems are considered intractable for classical computers but can be efficiently solved by quantum computers using algorithms like Shor's algorithm. Once quantum computers reach a sufficient scale and reliability, they could decrypt sensitive data, forge digital signatures, and compromise secure communications that rely on classical cryptographic techniques. The urgency of preparing for this quantum threat is now well recognized among researchers, governments, and cybersecurity professionals.

Post-quantum cryptography seeks to mitigate these threats by developing new cryptographic algorithms that can resist attacks by both classical and quantum computers. These algorithms are based on mathematical problems such as lattice-based, code-based, multivariate polynomial, and hash-based constructions. However, developing, standardizing, and deploying these algorithms on a global scale requires extensive collaboration between nations, institutions, and organizations. No single country or entity can address the challenges posed by quantum threats alone, especially when cyber systems and data flows transcend national borders.

International cooperation in cybersecurity related to PQC is vital for several reasons. First, cryptographic standards must be globally interoperable. The internet and other digital systems are inherently international in nature, and if countries adopt incompatible PQC algorithms, it could fragment communication protocols, reduce efficiency, and create vulnerabilities. For instance, secure internet communication protocols like TLS must function seamlessly across different devices, platforms, and networks worldwide. This requires agreement on which PQC algorithms to use and how to implement them securely. International bodies such as the International Organization for Standardization (ISO) and the Internet Engineering Task Force (IETF) play critical roles in developing and promoting these standards.

Second, the process of evaluating and vetting cryptographic algorithms benefits greatly from broad international participation. The U.S. National Institute of

Standards and Technology (NIST) has led the PQC standardization effort, but it has welcomed submissions and feedback from cryptographers around the world. This open and transparent process increases the robustness of the resulting standards by subjecting them to global scrutiny. Diverse expertise, testing in different environments, and attack simulations by international researchers help ensure that the algorithms are secure, efficient, and practical. This collaborative approach enhances trust and adoption across countries.

Third, global cooperation is necessary to ensure equitable access to quantum-resistant technologies. While technologically advanced countries may have the resources to develop and implement PQC rapidly, many developing nations face limitations in terms of funding, infrastructure, and technical capacity. Without coordinated support, a global digital divide may emerge in cybersecurity preparedness. International organizations, such as the United Nations' International Telecommunication Union (ITU) and the World Bank, have a role in facilitating capacity building and knowledge transfer. Collaborative projects, training programs, and funding mechanisms can help less-developed countries upgrade their cybersecurity infrastructure and adopt PQC solutions in time to address emerging threats.

Fourth, international agreements and policy harmonization are essential to manage the legal and regulatory implications of PQC. Data protection laws, export controls on cryptographic products, and requirements for government and defense communications vary across jurisdictions. Without harmonization, conflicting regulations could hinder the global deployment of PQC.

Bilateral and multilateral agreements can help align regulatory frameworks and encourage cooperation on issues such as algorithm certification, compliance standards, and secure cross-border data flows. Institutions like the European Union, the Organisation for Economic Co-operation and Development (OECD), and the G7 have been increasingly active in promoting dialogue and coordination on digital security policy.

Fifth, coordinated action is necessary to mitigate the threat of state-sponsored cyber operations and cybercrime exploiting quantum vulnerabilities. Cyberattacks are often carried out by actors based in one country targeting entities in another, making international law enforcement cooperation crucial. As quantum computing capabilities evolve, malicious actors may attempt to gain an advantage by harvesting encrypted data today in the hope of decrypting it later with quantum computers. Preventing this "store now, decrypt later" strategy requires global monitoring, intelligence sharing, and joint cybersecurity efforts. Institutions such as INTERPOL, Europol, and national cybersecurity centers must collaborate to detect, prevent, and respond to quantum-related cyber threats. Sixth, promoting responsible innovation and ethical use of quantum technologies is an international priority. As countries invest in quantum research, they must ensure that these technologies are developed and used in ways that uphold international norms and human rights. The development of international codes of conduct and best practices for quantum cybersecurity can help prevent misuse and ensure transparency. Additionally, countries can work together to establish frameworks for auditing and certifying quantum-secure systems, ensuring that they meet high ethical and security standards.

Seventh, fostering a global research ecosystem for PQC strengthens resilience and innovation. Collaborative research initiatives and academic partnerships enable the sharing of knowledge, tools, and datasets, accelerating the development and deployment of PQC. International conferences, workshops, and research grants can support the growth of a diverse and inclusive cryptographic community. Educational programs that involve cross-border collaboration can help cultivate the next generation of cryptographers, engineers, and policymakers with expertise in quantum-safe technologies.

Eighth, the role of the private sector in international cooperation on PQC cannot be overlooked. Technology companies, cloud providers, and financial institutions operate on a global scale and have a vested interest in securing their platforms against quantum threats. These organizations often drive the implementation of cryptographic changes and influence public policy. Through industry consortia and public-private partnerships, businesses can contribute to the standardization and deployment of PQC solutions. Sharing best practices, participating in interoperability testing, and engaging with regulators are important ways the private sector can support global efforts.

CHAPTER 10

Current Challenges and Future Directions

10.1 EFFICIENCY VS. SECURITY TRADE-OFFS

Post-quantum cryptography (PQC) has emerged as a critical field in the evolving landscape of cybersecurity. The advancement of quantum computing poses a significant threat to traditional cryptographic schemes such as RSA and elliptic curve cryptography. These systems are foundational to securing internet communications, financial transactions, government data, and more. Quantum computers, with their ability to efficiently solve certain mathematical problems, have the potential to render these classical algorithms obsolete. As a response, researchers have been developing quantum-resistant algorithms that could secure data even in the face of quantum adversaries.

Efficiency in cryptography generally refers to how quickly and resource-effectively cryptographic operations can be performed. This includes the time required to generate keys, sign or encrypt data, and verify signatures or decrypt messages, as well as the computational load and memory requirements.

Traditional public-key algorithms like RSA and ECC have enjoyed widespread deployment in part due to their relatively efficient performance, especially for key exchange and digital signature applications. They offer strong security with manageable computational costs and small key sizes. In contrast, many PQC algorithms require significantly more computational resources and bandwidth, potentially straining systems with limited processing power or memory, such as embedded devices or mobile platforms.

Security, in the context of PQC, means resistance to attacks from both classical and quantum computers. Since the arrival of quantum computers capable of breaking RSA and ECC is anticipated, the new cryptographic standards must offer long-term confidentiality and integrity, especially for sensitive or future-facing data. The security of PQC algorithms relies on problems like lattice-based assumptions, multivariate polynomial systems, code-based constructions, and hash-based signatures. These problems are believed to be hard even for quantum computers, but often at the expense of increased algorithmic complexity. This added complexity can result in higher latency, larger data exchanges, and slower processing times, which are all concerns for real-world applications where performance is crucial. The tension between efficiency and security becomes particularly evident when PQC algorithms are evaluated for standardization and practical deployment. NIST's Post-Quantum Cryptography Standardization Project has highlighted this dynamic, as candidate algorithms are assessed not only for their theoretical security but also for their practical feasibility.

For instance, lattice-based schemes like Kyber (for key encapsulation) and Dilithium (for digital signatures) have shown strong promise and have been selected for standardization. However, they still present challenges in terms of implementation, especially in constrained environments such as IoT devices or smart cards. The key and ciphertext sizes for Kyber are notably larger than those of RSA or ECC, and Dilithium's signatures are significantly longer, which impacts bandwidth and storage.

One of the major efficiency concerns in PQC is communication overhead. Many PQC schemes produce keys and ciphertexts that are several times larger than those in classical cryptography. In systems that rely on constrained bandwidth, such as satellite communication or low-power wireless networks, this can lead to transmission delays and increased costs. Moreover, the longer processing times may not be acceptable in latency-sensitive applications, such as real-time financial trading, where milliseconds can make a significant difference. Ensuring a smooth transition to PQC will require careful optimization and, in some cases, architectural changes to accommodate these demands.

Another area where the efficiency-security trade-off is apparent is in hybrid cryptographic systems. In the transitional period before quantum computers become practically available, many organizations are adopting hybrid models that combine classical and post-quantum algorithms. These models aim to provide both immediate security and future resilience. While hybrid schemes improve confidence in long-term security, they also increase the computational burden, as systems must perform double the cryptographic operations. This additional overhead can slow down transactions, increase energy consumption, and complicate protocol design. Nonetheless, for applications where security cannot be compromised, such as government communications or critical infrastructure, these trade-offs may be deemed acceptable.

Deployment constraints also shape the efficiency-security balance. Many existing systems were designed with classical cryptography in mind, and retrofitting them to support PQC can be challenging. Software updates, hardware modifications, and changes to network protocols may be required to handle the increased data sizes and processing requirements. This is particularly problematic for legacy systems that are no longer actively maintained or that operate in highly regulated environments. In such cases, implementers must evaluate whether the increased security provided by PQC justifies the cost and complexity of upgrading systems, or whether alternative protective measures can be employed temporarily.

In the cloud computing and enterprise space, efficiency considerations are crucial due to the scale of operations. Cloud providers must perform millions of secure transactions per second and store vast amounts of encrypted data. PQC algorithms that significantly increase processing times or storage requirements could impact service quality and operational costs. Therefore, cloud service providers are actively involved in PQC research and benchmarking, aiming to identify algorithms that provide the best trade-off between strong security guarantees and manageable performance implications.

Despite these challenges, there is significant ongoing research aimed at improving the efficiency of PQC algorithms. Optimizations at the algorithmic, software, and hardware levels are being explored. For instance, specialized hardware accelerators can help mitigate the performance penalties of computationally intensive PQC operations. Software libraries are being optimized to leverage modern processor instructions and parallel computing capabilities. Additionally, lightweight post-quantum schemes are being developed specifically for resource-constrained environments, aiming to deliver adequate security with reduced computational demands.

Moreover, different applications have varying tolerance for performance penalties, allowing implementers to choose algorithms tailored to specific use cases. High-security environments, such as national defense and financial sectors, may prioritize security over efficiency, accepting higher overheads for stronger protection. Conversely, consumer applications like messaging apps or e-commerce platforms might favor algorithms with balanced performance characteristics to ensure a smooth user experience.

The development of PQC also brings to the forefront the importance of formal verification and side-channel resistance. Strong theoretical security does not guarantee real-world safety, as implementation flaws and side-channel attacks can undermine even the most secure algorithms. Ensuring that PQC algorithms are not only mathematically sound but also resistant to timing attacks, power analysis, and other implementation-based vulnerabilities adds further layers of complexity and potential performance trade-offs.

10.2 POST-PQC THREAT MODELS (E.G., QUANTUM-RESISTANT YET VULNERABLE TO SIDE-CHANNELS)

The advent of post-quantum cryptography marks a pivotal moment in cybersecurity, but its success depends on addressing more than just quantum threats. Side-channel attacks, fault injections, and implementation flaws remain potent dangers that can undermine even the most mathematically secure PQC systems. A comprehensive security strategy must integrate algorithmic strength with side-channel resilience, robust standardization with rigorous certification, and theoretical advances with practical safeguards. By adopting a holistic approach—one that considers both quantum and classical threat models—organizations can ensure that their transition to post-quantum cryptography truly enhances security rather than introducing new vulnerabilities. The race to quantum resistance is not just about surviving the future of computing; it's about building systems that are secure, resilient, and trustworthy in all dimensions.

The future of post-quantum cryptography lies in bridging the gap between mathematical robustness and practical security. Research must focus on developing inherently side-channel-resistant PQC algorithms, perhaps leveraging techniques like fully homomorphic encryption (FHE) or secure multi-party computation (MPC) to minimize leakage. Automated tools for side-channel analysis and verification can help developers identify and patch vulnerabilities early in the design phase. Additionally, the integration of artificial intelligence for anomaly detection could enhance real-time defenses against emerging side-channel threats. As quantum computing becomes a reality, the cybersecurity community must adopt a dual focus: defending against quantum attacks while simultaneously hardening systems against classical exploitation.

10.2.1 Side-Channel Attacks

Side-channel attacks have long plagued classical cryptographic systems, and their relevance persists in the post-quantum era. Unlike brute-force or algorithmic cryptanalysis, side-channel attacks bypass theoretical security by observing unintended information leakage from a device's operation. For example, a power analysis attack on a lattice-based PQC implementation could reveal secret keys by measuring fluctuations in energy usage during cryptographic operations.

Similarly, timing attacks might exploit delays in modular arithmetic computations to infer sensitive data. The structural complexity of many PQC algorithms—such as those based on lattices, hash functions, or multivariate polynomials—can introduce new side-channel risks. Larger key sizes, intricate mathematical operations, and memory-intensive computations may inadvertently expose more attack surfaces, making PQC implementations a fertile ground for side-channel exploitation.

10.2.2 Role of Implementation Security

The security of post-quantum cryptography hinges not only on algorithm selection but also on implementation quality. A theoretically quantum-resistant algorithm can still fail catastrophically if deployed without side-channel protections. Secure implementation practices include constant-time programming to eliminate timing leaks, masking techniques to obscure power consumption patterns, and isolation mechanisms to prevent electromagnetic or acoustic eavesdropping.

Hardware-based solutions, such as trusted execution environments (TEEs) and hardware security modules (HSMs), can further mitigate risks by physically shielding cryptographic operations. Standardization bodies like NIST have begun addressing these concerns by incorporating implementation security into PQC evaluation criteria. However, the responsibility ultimately falls on developers and system integrators to ensure that PQC deployments are resilient against both quantum and classical adversaries.

10.2.3 Emerging Threats

Beyond traditional side-channel methods, fault injection attacks pose a growing threat to post-quantum cryptographic systems. These attacks involve deliberately inducing computational errors—via voltage glitching, laser injection, or electromagnetic pulses—to corrupt cryptographic operations and reveal secret information. Many PQC algorithms, particularly those with complex error-correcting mechanisms or rejection sampling steps, may exhibit exploitable behavior under fault conditions.

For instance, a successful fault attack on a lattice-based key exchange could force the generation of weak keys, enabling later decryption. Defending against such attacks requires a multi-layered approach, combining algorithmic robustness (e.g., error detection and redundancy) with hardware-level protections (e.g., sensors for environmental manipulation). As PQC moves into critical infrastructure, fault resistance must become a core consideration in both design and deployment.

10.2.4 Standardization and Certification Challenges

The standardization of post-quantum cryptography is not limited to mathematical security—it must also encompass implementation guidelines to thwart side-channel and physical attacks. Current PQC standardization efforts, such as NIST's ongoing process, primarily focus on algorithmic strength and performance, leaving implementation security as a secondary concern. This gap could lead to certified PQC systems that are theoretically quantum-resistant but practically vulnerable.

To address this, certification frameworks like Common Criteria and FIPS 140-3 must evolve to include rigorous side-channel testing for PQC modules. International collaboration among standardization bodies, industry groups, and academic researchers will be essential to develop comprehensive security criteria that cover both quantum and classical threat models. Without such measures, widespread PQC adoption could inadvertently introduce new security weaknesses.

10.2.5 Mitigation Strategies for Side-Channel-Resistant PQC

Protecting post-quantum cryptographic systems from side-channel attacks demands a proactive and multi-faceted approach. Algorithmic countermeasures, such as masking and blinding techniques, can obscure sensitive computations from attackers. Hardware-based mitigations, including secure enclaves and tamper-resistant chips, provide physical barriers against leakage. Software defenses, such as compiler-level optimizations for constant-time execution, reduce the risk of timing vulnerabilities.

Additionally, continuous monitoring and anomaly detection can identify potential side-channel exploitation in real-time. A critical best practice is the principle of defense-in-depth, where multiple layers of protection are employed to ensure that a breach in one does not compromise the entire system. As PQC algorithms mature, their accompanying security standards must prioritize these mitigations to prevent quantum-resistant systems from falling prey to classical attacks.

10.2.6 Human Factor: Training and Awareness in PQC Security

Even the most secure PQC implementations can be undermined by human error or insufficient expertise. Developers, system administrators, and cybersecurity professionals must be educated on the unique challenges of post-quantum cryptography, including side-channel risks.

Training programs should emphasize secure coding practices for PQC, threat modeling for hybrid systems, and incident response for side-channel breaches. Organizations transitioning to PQC must also foster collaboration between cryptographers, hardware engineers, and security practitioners to ensure holistic protection. Furthermore, policymakers and industry leaders should advocate for funding and research into side-channel-resistant PQC, recognizing that theoretical security alone is insufficient for real-world deployment.

10.2.7 Case Studies

Several post-quantum cryptographic schemes have already demonstrated susceptibility to side-channel attacks, underscoring the need for robust countermeasures. Lattice-based algorithms, such as Kyber and Dilithium, rely on polynomial multiplications and modular reductions, which can leak timing information if not implemented carefully. Research has shown that even masked implementations of these algorithms can be compromised through advanced differential power analysis (DPA). Hash-based signatures, like SPHINCS+, face challenges in random number generation, where poor entropy management can lead to key recovery through fault injection. Code-based cryptography, exemplified by Classic McEliece, may be vulnerable to cache-timing attacks due to its reliance on large lookup tables. These examples illustrate that while PQC algorithms are mathematically secure against quantum computers, their real-world deployments must be rigorously hardened against side-channel threats.

10.3 QUANTUM KEY DISTRIBUTION (QKD) VS. PQC

The emergence of quantum computing has necessitated the development of two fundamentally different approaches to quantum-resistant security: Quantum Key Distribution (QKD) and Post-Quantum Cryptography (PQC). These paradigms represent distinct philosophical and technical responses to the quantum threat, each with its own advantages, limitations, and appropriate use cases. As organizations prepare for the quantum computing era, understanding the differences between these approaches becomes essential for developing comprehensive security strategies. The complementary nature of QKD and PQC suggests that rather than being competing solutions, they may represent different tools in a broader quantum-safe security toolkit.

The importance of comparing these approaches stems from their fundamentally different security guarantees and implementation characteristics. QKD provides information-theoretic security based on the laws of quantum physics, offering provable security guarantees that differ from the computational hardness assumptions underlying PQC. This distinction has significant implications for security architecture, implementation complexity, and operational requirements. As the cryptographic community evaluates these approaches, the potential for synergistic combinations that leverage the strengths of both paradigms becomes increasingly apparent. The transition to quantum-safe security will likely involve strategic decisions about where and how to deploy each approach based on specific security requirements and operational constraints.

10.3.1 Security Guarantees and Threat Models

The security guarantees provided by QKD and PQC differ fundamentally in their nature and the threat models they address. QKD offers information-theoretic security based on physical principles, where the security of the key exchange process is guaranteed by the laws of quantum mechanics rather than computational assumptions. This provides a level of security that is theoretically unbreakable, assuming the physical implementation is correct and the quantum channel is properly secured. The threat model for QKD focuses primarily on the physical layer, where attacks might attempt to intercept or manipulate quantum signals, with security guarantees holding as long as these physical protections remain intact.

PQC, by contrast, provides computational security guarantees that depend on the assumed hardness of specific mathematical problems. While these problems are believed to be resistant to quantum attacks, their security remains based on computational assumptions rather than physical laws. The threat model for PQC is more similar to classical cryptography, focusing on algorithmic attacks and implementation vulnerabilities. The security of PQC implementations depends on proper parameter selection, correct implementation, and the continued validity of underlying hardness assumptions. This creates a different security posture than QKD, with implications for trust models, key management, and long-term security planning.

10.3.2 Implementation Complexity and Deployment Challenges

The implementation complexity and deployment challenges of QKD and PQC differ significantly due to their fundamentally different technological requirements. QKD systems require specialized hardware capable of generating, transmitting, and measuring quantum states, typically involving single-photon sources and detectors. The deployment of QKD networks necessitates dedicated quantum channels, often through fiber optic cables or free-space optical links, creating infrastructure requirements that can limit deployment scenarios. The physical nature of QKD also introduces distance limitations, as quantum signals cannot be amplified like classical signals, requiring trusted nodes or quantum repeaters for long-distance communication.

PQC implementations, by contrast, can leverage existing computational infrastructure, making them potentially easier to deploy in many scenarios. The primary implementation challenges for PQC involve the larger key sizes and more complex operations characteristic of many quantum-resistant algorithms.

These can require more computational resources and memory than classical cryptographic primitives, potentially creating performance bottlenecks in some systems. The deployment of PQC typically involves software updates and algorithm replacements rather than new hardware installations, though some applications may benefit from hardware acceleration for performance-critical operations. The migration to PQC also requires careful planning to ensure compatibility with existing systems and protocols during transition periods.

10.3.3 Performance Characteristics and Operational Considerations

The performance characteristics of QKD and PQC implementations present different operational considerations that influence their suitability for various applications. QKD systems typically have lower key generation rates compared to computational cryptographic methods, with current implementations often producing keys at rates measured in kilobits per second. This can limit their applicability for high-bandwidth communications unless combined with classical encryption methods that use the QKD-generated keys. The physical infrastructure requirements of QKD also create operational constraints, as quantum channels must be carefully maintained and protected to ensure security guarantees.

PQC implementations generally offer higher key generation rates than QKD but may require more computational resources than classical cryptographic algorithms. The performance characteristics vary significantly between different PQC algorithm families, with some schemes offering better throughput or latency characteristics than others. The operational considerations for PQC include the need for careful parameter selection to balance security and performance, as well as potential requirements for hardware acceleration in performance-critical applications. The migration to PQC also involves considerations of algorithm agility, where systems may need to support multiple cryptographic schemes during transition periods.

10.3.4 Use Cases and Application Scenarios

The distinct characteristics of QKD and PQC make them suitable for different application scenarios based on specific security requirements and operational constraints. QKD finds particular applicability in scenarios where long-term security guarantees are paramount, such as protecting classified government communications or securing critical infrastructure with extended operational lifetimes. The information-theoretic security of QKD makes it attractive for applications where the potential for future cryptanalytic advances poses unacceptable risks. High-security government facilities, financial institutions with long-term data protection needs, and critical infrastructure operators represent typical early adopters of QKD technology.

PQC, with its broader applicability and easier deployment characteristics, finds use in scenarios where computational cryptography remains appropriate but quantum resistance is required. This includes most general-purpose secure communications, data protection systems, and authentication mechanisms where the infrastructure requirements of QKD would be impractical.

Enterprise networks, cloud services, and consumer applications represent typical use cases for PQC implementations. The ability to deploy PQC through software updates makes it particularly suitable for existing systems that require migration to quantum-resistant algorithms. Some applications may benefit from hybrid approaches that combine elements of both QKD and PQC to leverage their respective strengths.

10.3.5 Standardization and Interoperability Considerations

The standardization landscape for QKD and PQC presents different challenges and opportunities due to their distinct technological characteristics. QKD standardization efforts focus primarily on physical layer specifications and protocol definitions that ensure interoperability between different vendor implementations. The ETSI Quantum Key Distribution Industry Specification Group and the ITU-T Study Group 13 have developed standards for QKD systems that address both point-to-point and networked implementations. These standards help ensure that QKD equipment from different manufacturers can work together while maintaining security guarantees.

PQC standardization has followed a more traditional cryptographic path through efforts like the NIST Post-Quantum Cryptography Standardization Project. This process has evaluated multiple algorithm candidates across different mathematical families, selecting those believed to offer the best combination of security and performance characteristics. The standardization of PQC involves not just algorithm specifications but also implementation guidelines that address potential vulnerabilities and ensure consistent security levels. The interoperability considerations for PQC focus on ensuring that different implementations can communicate securely while supporting algorithm agility during transition periods.

10.4 FULLY HOMOMORPHIC ENCRYPTION IN A PQC WORLD

The evolution of cryptography has been driven by the continuous demand for more secure, efficient, and privacy-preserving computational techniques. One of the most transformative innovations in recent decades is Fully Homomorphic Encryption (FHE), a cryptographic method that allows computation on encrypted data without needing to decrypt it first. FHE enables a wide range of secure applications such as privacy-preserving cloud computing, secure machine learning, and encrypted database queries. However, as the rise of quantum computing threatens many existing cryptographic schemes, it has become essential to explore the resilience of FHE systems in a post-quantum cryptographic (PQC) world.

FHE operates on the principle that one can perform arithmetic or logical operations directly on ciphertexts, and the result, when decrypted, matches the result of operations as if they had been performed on the plaintexts. This unique property opens up immense possibilities in fields where data privacy is paramount. A medical researcher can analyze encrypted patient data without seeing the actual information.

Financial institutions can process confidential client information while preserving privacy. However, the algorithms behind FHE are computationally intensive, and the implementation of FHE schemes based on traditional cryptographic assumptions is already considered resource-demanding. With the threat of quantum computers on the horizon, ensuring that FHE also resists quantum attacks adds another layer of complexity.

The classical security of many FHE schemes is based on hard mathematical problems such as factoring large integers or computing discrete logarithms. These problems can be solved efficiently using quantum algorithms like Shor's algorithm, rendering such schemes insecure in a quantum computing context. As a result, FHE schemes must be constructed from quantum-resistant assumptions. Fortunately, many of the most promising FHE systems today are based on lattice-based cryptography, specifically on problems like the Learning With Errors (LWE) or Ring Learning With Errors (Ring-LWE), which are believed to be secure against quantum adversaries. These problems remain hard even for quantum computers, making them suitable foundations for PQC-compatible FHE systems.

The most widely known and researched FHE schemes in the PQC context are those developed from lattice-based constructions. Lattice-based cryptography offers several advantages, including simplicity of construction, provable security under worst-case assumptions, and flexibility in designing encryption schemes.

Notably, schemes such as the Brakerski-Gentry-Vaikuntanathan (BGV), Fan-Vercauteren (FV), and Cheon-Kim-Kim-Song (CKKS) rely on LWE or its variants and provide different trade-offs in terms of precision, performance, and support for arithmetic operations. These schemes are already being used in post-quantum FHE implementations, making lattice-based FHE the primary candidate for secure encrypted computing in the quantum era.

Despite the theoretical soundness of lattice-based FHE, one of the main challenges lies in efficiency. Even without the quantum threat, FHE is known for its high computational overhead. Homomorphic operations are slower than their plaintext counterparts by several orders of magnitude. The addition of post-quantum requirements often exacerbates this problem. FHE ciphertexts in lattice-based schemes are typically large, and homomorphic operations require expensive polynomial multiplications, noise management, and bootstrapping techniques to keep ciphertexts usable. These processes consume significant memory and computational resources, which can become barriers to real-world deployment, particularly in resource-constrained environments like mobile devices or IoT systems.

Bootstrapping is one of the most resource-intensive aspects of FHE and is critical for enabling unlimited homomorphic operations. This process refreshes the ciphertext to reduce accumulated noise and ensures continued correctness of computation. In the PQC context, bootstrapping must be compatible with quantum-safe cryptographic primitives. Recent advancements have improved bootstrapping efficiency, but challenges remain in achieving acceptable latency for practical applications. Research efforts continue to focus on optimizing parameters, reducing ciphertext size, and leveraging hardware acceleration to make post-quantum FHE more viable.

In practical applications, FHE combined with post-quantum assumptions opens doors to secure cloud computing and privacy-preserving AI that can withstand both classical and quantum attacks. For instance, a hospital could outsource the computation of patient diagnostics to a cloud service while maintaining strict confidentiality under encryption, without fear of quantum decryption. Financial analysis and fraud detection algorithms could be run on encrypted data, ensuring customer privacy and regulatory compliance even in a future quantum-powered cyber environment. Governments could securely outsource analytics on encrypted census or intelligence data, confident that the encryption remains secure even if adversaries develop quantum capabilities.

In a PQC world, the use of FHE is not only a matter of privacy but also of sovereignty and resilience. As nations and organizations prepare for quantum threats, FHE offers a robust solution for protecting sensitive computations. It allows data owners to retain control over their information, delegating processing without exposing the raw content. This characteristic is particularly important for countries seeking to ensure data sovereignty and reduce dependency on foreign cloud providers, a concern amplified by growing geopolitical tensions and cybersecurity threats.

Moreover, integrating FHE with other post-quantum cryptographic primitives is a key direction for future development. Hybrid approaches combining FHE with post-quantum secure key exchange protocols, authentication mechanisms, and secure multiparty computation can create comprehensive systems that are secure against quantum threats across all layers of communication and computation. This integration requires careful protocol design to ensure consistency, efficiency, and interoperability, which remains an ongoing area of research and standardization.

Standardization efforts are also essential for the adoption of FHE in the PQC landscape. While NIST's PQC standardization process primarily focuses on key exchange and digital signatures, other organizations such as ISO, ITU, and industry consortia are beginning to explore standardized frameworks for FHE. Interoperability standards, performance benchmarks, and verified implementations are needed to foster trust and facilitate widespread deployment. Open-source libraries such as Microsoft SEAL, IBM HELib, and PALISADE are contributing significantly to this effort by providing developers and researchers with tools to experiment with and evaluate post-quantum FHE systems.

Another crucial dimension is the legal and ethical implications of FHE in a post-quantum world. As encryption becomes more powerful, it may raise concerns about lawful access, surveillance, and regulatory oversight. Governments may face new challenges in balancing citizens' rights to privacy with the need for national security. The debate over "backdoors" and lawful interception in cryptographic systems is likely to intensify, especially when encryption schemes become virtually unbreakable even by quantum computers. Therefore, alongside technological development, there must be a transparent and inclusive dialogue among stakeholders, including governments, industry, academia, and civil society, to establish norms and frameworks for responsible use of post-quantum FHE.

10.5 PREPARING FOR CRYPTOGRAPHIC DISRUPTION AND INNOVATION CYCLES

The shift to post-quantum cryptography represents both a significant challenge and an opportunity to build more resilient security infrastructures. By viewing PQC implementation not as a destination but as part of an ongoing process of cryptographic evolution, organizations can position themselves to adapt more effectively to future disruptions. This requires moving beyond tactical algorithm replacement to develop strategic capabilities in cryptographic lifecycle management, system agility, and transition planning. The organizations that will thrive in this environment are those that recognize cryptographic resilience as a competitive advantage and core operational requirement. As we stand on the brink of the quantum computing era, the lessons learned from preparing for PQC will serve as valuable foundations for navigating whatever cryptographic challenges emerge next in our increasingly digital world.

10.5.1 Cryptographic Lifecycles and Obsolescence

Every cryptographic algorithm has a finite lifespan, determined by advances in computing power, cryptanalysis techniques, and mathematical breakthroughs. The history of cryptography demonstrates this pattern clearly, with once-secure algorithms like RSA-512 and MD5 becoming obsolete within decades of their introduction. In the quantum era, this lifecycle is accelerating, as demonstrated by NIST's ongoing PQC standardization process which anticipates the need for regular updates to cryptographic standards. Organizations must develop strategies for cryptographic lifecycle management that account for this accelerated pace of change. This includes establishing processes for continuous monitoring of cryptographic vulnerabilities, maintaining inventories of cryptographic assets, and planning for orderly transitions before algorithms become compromised. The concept of "cryptographic shelf life" becomes particularly important in this context, with security teams needing to anticipate not just current threats but also future vulnerabilities that might emerge during a system's operational lifetime.

10.5.2 Building Cryptographic Agility into Systems Architecture

The fundamental lesson from previous cryptographic transitions is that systems must be designed from the outset to accommodate change. Cryptographic agility - the ability to swap out cryptographic primitives without requiring complete system redesign - should be a core architectural principle for all security-sensitive systems. This requires moving away from hard-coded cryptographic implementations toward modular, parameterized designs that can easily incorporate new algorithms and key lengths. Modern cryptographic frameworks are increasingly adopting this approach, with features like algorithm negotiation in TLS 1.3 providing a model for how systems can maintain interoperability while supporting multiple cryptographic options. However, true cryptographic agility extends beyond just algorithm selection to encompass key management, protocol design, and system integration. Organizations should implement standardized cryptographic interfaces, maintain comprehensive algorithm transition plans, and ensure that their systems can support hybrid cryptographic modes during transition periods.

10.5.3 Challenge of Cryptographic Inventory and Asset Management

One of the most significant barriers to effective cryptographic transition is the lack of visibility into where and how cryptography is used across complex IT ecosystems. Large organizations may have thousands of systems, devices, and applications that rely on cryptographic functions, often implemented in different ways and with varying levels of security. Preparing for post-quantum cryptography requires first establishing a complete cryptographic inventory - identifying all systems that use cryptography, the specific algorithms they employ, and their dependencies on particular cryptographic properties. This inventory process is complicated by the pervasive nature of cryptography in modern computing, from obvious uses like SSL/TLS to more obscure applications in firmware, database indices, and system logging. Automated discovery tools and cryptographic auditing techniques can help organizations map their cryptographic footprint, but this remains a challenging task that requires ongoing attention as systems evolve.

10.5.4 Hybrid Cryptography as a Transition Strategy

The transition to post-quantum cryptography cannot happen instantaneously across all systems. Hybrid cryptography - the simultaneous use of classical and quantum-resistant algorithms - offers a practical transition strategy that maintains security while allowing for gradual migration. In a hybrid approach, systems might use both RSA and a lattice-based algorithm for digital signatures, or combine ECC with a code-based key exchange mechanism. This provides defense in depth during the transition period, as an attacker would need to break both cryptographic systems to compromise security. However, hybrid cryptography introduces its own challenges, including increased computational overhead, larger key sizes, and more complex protocol negotiations. Organizations implementing hybrid solutions must carefully consider performance implications, interoperability requirements, and the eventual process of deprecating the classical components once the transition is complete.

10.5.5 Standardization and Interoperability Challenges

The global nature of digital infrastructure means that cryptographic transitions must be coordinated across industries and national boundaries. The current PQC standardization process led by NIST is a critical step in this direction, but standardization alone doesn't guarantee smooth transitions. Different industries and jurisdictions may adopt new standards at different paces, leading to potential interoperability issues during transition periods. Some sectors may face regulatory requirements to implement PQC earlier than others, while legacy systems in critical infrastructure may resist timely upgrades. The development of transition frameworks and testing programs can help mitigate these challenges, but organizations must be prepared to manage periods of cryptographic heterogeneity where different parts of their ecosystem operate at different levels of cryptographic security.

10.5.6 Human Factor in Cryptographic Transitions

Technological solutions alone cannot ensure successful cryptographic transitions - organizational and human factors play equally critical roles. Cryptographic migrations require coordination across multiple teams including security architects, developers, operations staff, and business stakeholders. Many organizations lack sufficient in-house expertise in advanced cryptography, creating knowledge gaps that can hinder effective transition planning. Training programs and workforce development initiatives are essential to building the necessary skills for PQC implementation and management. Additionally, change management processes must address the cultural resistance that often accompanies major technological shifts, particularly when they require modifications to long-established security practices. Executive leadership must understand both the strategic importance of cryptographic readiness and the resource commitments required to achieve it.

10.5.7 Preparing for the Next Cryptographic Disruption

While post-quantum cryptography currently dominates discussions about cryptographic futures, it represents just one point in an ongoing cycle of disruption and innovation. The security community must look beyond the immediate PQC transition to prepare for future cryptographic challenges that may emerge from developments in areas like artificial intelligence, biological computing, or unforeseen mathematical breakthroughs.

This requires establishing institutional processes for continuous cryptographic evaluation, maintaining research partnerships with academia and industry, and fostering a culture of cryptographic awareness throughout organizations. Investment in fundamental cryptographic research and development of next-generation security primitives will be essential to staying ahead of emerging threats. Perhaps most importantly, organizations must internalize that cryptographic transition is not a one-time project but an ongoing capability that must be sustained as part of core security operations.

CHAPTER 11

Case Studies and Sectoral Adoption

11.1 GOOGLE'S CECPQ AND HYBRID TLS EXPERIMENTS

In the evolving landscape of digital security, the emergence of quantum computing presents a significant threat to classical cryptographic systems. Recognizing this future challenge, Google has taken a proactive role in preparing for a post-quantum world by initiating real-world experiments to test quantum-resistant cryptographic algorithms. One of the pioneering efforts in this domain is Google's CECPQ (Combined Elliptic Curve and Post-Quantum) project, which involved the deployment of hybrid cryptographic key exchange mechanisms within the widely used Transport Layer Security (TLS) protocol.

The CECPQ experiment is centered around hybrid cryptography, an approach that combines traditional, well-established cryptographic algorithms with experimental post-quantum algorithms to ensure security both now and in the future. In CECPQ1, Google integrated the New Hope key exchange algorithm—a post-quantum lattice-based algorithm based on the Ring Learning With Errors (Ring-LWE) problem—with the existing elliptic curve key exchange mechanism (X25519). By doing so, even if the post-quantum component were later found to be insecure or inefficient, the classical component would continue to provide a robust level of protection. Conversely, in the event of a successful quantum attack on classical elliptic curve cryptography, the post-quantum component would still ensure data security.

The deployment of CECPQ1 occurred in Google's Chrome browser and was tested on connections to Google servers, enabling the team to collect valuable real-world data. This approach allowed Google to study the integration of post-quantum algorithms in a manner that was invisible to users and non-disruptive to services.

One of the key challenges addressed in the experiment was maintaining compatibility with existing infrastructure while evaluating the overhead in bandwidth and latency introduced by larger key sizes and more complex computations. The CECPQ1 experiment concluded successfully and demonstrated that hybrid key exchange could be practically deployed at scale, with manageable trade-offs in performance. Following CECPQ1, Google initiated CECPQ2, which explored a new post-quantum algorithm known as HRSS (Hülsing, Rijneveld, Schanck, Schwabe)—a variant of the LWE problem—combined with X25519. CECPQ2 focused on reducing latency and improving efficiency compared to CECPQ1, and it represented a more refined approach as post-quantum cryptography standards evolved. In CECPQ2, the post-quantum portion was chosen to have a smaller ciphertext size while maintaining quantum resistance. This refinement responded to concerns raised during CECPQ1 regarding the bandwidth impact of large post-quantum key sizes, especially in mobile and low-latency environments.

Google's hybrid TLS experiments have played a foundational role in informing the global cryptographic community, including standardization bodies such as NIST and the IETF. These experiments demonstrated the practical considerations of deploying post-quantum cryptography, such as backward compatibility, handshake timings, and resource consumption. They provided empirical evidence that hybrid key exchange mechanisms can be deployed in real-world applications without breaking existing systems or sacrificing user experience.

11.2 CLOUDFLARE'S PQC TESTING AND DEPLOYMENT

As quantum computing continues to advance, the cryptographic foundations that underpin modern internet security face potential obsolescence. To address this looming threat, major technology companies have started implementing and testing quantum-resistant algorithms.

Among these, Cloudflare has been a leader in the experimentation and deployment of post-quantum cryptography (PQC), especially in the context of securing Internet traffic via TLS (Transport Layer Security). Cloudflare's approach to PQC focuses on real-world testing, collaboration with academic and industry partners, and supporting the broader effort to transition the internet infrastructure toward post-quantum security.

Cloudflare's primary focus in PQC has been the integration of hybrid key exchange algorithms within TLS, combining classical and post-quantum schemes to ensure forward security both now and in the quantum future. The hybrid model allows systems to rely on existing, well-tested cryptographic protocols while incorporating post-quantum components that can safeguard against future quantum adversaries. This dual approach not only preserves backward compatibility but also provides a defense-in-depth strategy where at least one part of the hybrid scheme must be broken to compromise the security of a session.

In 2019, Cloudflare began a large-scale experiment with Google and Mozilla to implement post-quantum key exchange algorithms in real-world internet traffic. One of the most significant outcomes of these experiments was the deployment of TLS connections using the post-quantum algorithm SIDH (Supersingular Isogeny

Diffie-Hellman), which was tested alongside traditional elliptic curve cryptography. This allowed Cloudflare to gather performance data on latency, handshake time, and resource consumption while measuring the impact on end-user experience. These trials proved essential in identifying the trade-offs involved in deploying PQC, such as larger key sizes and slower handshake performance, which could affect mobile or low-bandwidth users.

Cloudflare's post-quantum efforts expanded in tandem with the National Institute of Standards and Technology (NIST)'s PQC standardization project. As NIST narrowed down candidate algorithms for key exchange and digital signatures, Cloudflare began testing schemes like Kyber, a lattice-based key encapsulation mechanism that eventually emerged as one of NIST's finalists.

Cloudflare integrated Kyber into their infrastructure as part of hybrid TLS handshakes, enabling compatibility with both legacy and future systems. This deployment, although experimental at first, helped prove the viability of Kyber in high-throughput, low-latency environments.

Another major step by Cloudflare was the open-source release of their PQC tools and libraries. By providing implementations and technical documentation, Cloudflare enabled researchers, developers, and enterprises to experiment with and evaluate post-quantum cryptography in their own systems. These contributions have supported community-driven development and allowed stakeholders to prepare for a future cryptographic transition without waiting for formal mandates.

11.3 PQC IN FINANCIAL SYSTEMS AND DIGITAL BANKING

The financial sector has always placed significant importance on security and data integrity due to the sensitive nature of transactions and the constant threat of cyberattacks. With the emergence of quantum computing, traditional public-key cryptographic methods such as RSA and elliptic-curve cryptography are facing an existential threat.

These methods, currently foundational to digital banking and financial systems, are vulnerable to being broken by quantum algorithms like Shor's algorithm. As a result, Post-Quantum Cryptography (PQC) has become a critical area of focus for financial institutions, governments, and cybersecurity experts aiming to protect digital assets and infrastructures in the coming quantum era.

Financial systems rely heavily on cryptographic protocols to perform key operations such as user authentication, digital signatures, secure communications, and data confidentiality. Online banking transactions, mobile payments, ATM communications, and interbank transfers all utilize encryption protocols that could be compromised by sufficiently powerful quantum computers. This looming vulnerability has made the integration of PQC algorithms into financial systems a high priority. PQC algorithms are designed to resist attacks from both classical and quantum adversaries, ensuring long-term security even when quantum capabilities become mainstream.

Transitioning to PQC in financial systems presents multiple challenges, both technical and operational. Financial institutions must ensure that the new cryptographic schemes can integrate seamlessly with existing payment processing systems, secure APIs, and digital certificate infrastructures. Unlike some sectors where new technologies can be adopted rapidly, banking systems operate under stringent regulatory frameworks and high reliability requirements, which makes system-wide cryptographic upgrades complex and risk-sensitive. Any transition must guarantee backward compatibility and minimal disruption to daily operations.

One of the initial approaches toward integrating PQC in financial services involves hybrid encryption models. These models combine current classical encryption algorithms with quantum-resistant algorithms in a single cryptographic protocol. This ensures that even if one part of the encryption scheme becomes vulnerable, the other continues to provide a level of security. This method allows financial institutions to begin deploying quantum-resistant elements without fully replacing existing cryptographic systems, thus supporting a gradual and secure transition.

Post-quantum digital signature schemes are also being evaluated for applications in secure document verification, transaction authorization, and identity validation in digital banking platforms. Financial institutions routinely use digital signatures to authenticate users, authorize fund transfers, and ensure transaction non-repudiation. Implementing PQC-based digital signatures can future-proof these essential functions against quantum threats. Lattice-based schemes like Dilithium and hash-based schemes such as SPHINCS+ have shown promise in offering strong quantum resistance while maintaining operational efficiency suitable for financial systems.

PQC deployment in banking also demands careful attention to hardware limitations, especially in mobile banking applications. The size of keys and the computational overhead of PQC algorithms must be optimized to prevent negative impacts on user experience. Furthermore, regulators and financial authorities need to establish guidelines and standards for PQC implementation across different types of financial entities to ensure consistent security practices industry-wide.

11.4 NATIONAL CRYPTOGRAPHIC TRANSITION PLANS (E.G., USA, EU, JAPAN)

As quantum computing moves from theoretical potential to practical development, nations across the globe have started formulating cryptographic transition plans to prepare for a post-quantum world. These transition plans aim to replace vulnerable classical encryption systems with quantum-resistant algorithms known as post-quantum cryptography (PQC). Countries like the United States, the European Union member states, and Japan are leading this movement by designing national strategies, setting standards, and encouraging widespread adoption of PQC within government, industry, and critical infrastructure sectors.

The United States has taken a proactive stance through the National Institute of Standards and Technology (NIST), which began the Post-Quantum Cryptography Standardization Project in 2016. This initiative is central to the U.S. cryptographic transition strategy. The project has focused on evaluating and selecting quantum-resistant cryptographic algorithms suitable for standardized use across federal agencies and commercial systems.

By 2022, NIST announced its initial selections for standardization, including the key encapsulation mechanism Kyber and digital signature algorithms like Dilithium and SPHINCS+. Building upon NIST's work, the White House released a National Security Memorandum in 2022 outlining a phased plan for federal agencies to inventory cryptographic systems, prioritize assets based on risk, and prepare for PQC integration. This memo emphasized early preparation and mandated timelines to begin the adoption of NIST-approved algorithms, reflecting the urgency placed on national cybersecurity in the quantum era.

In the European Union, the transition to post-quantum cryptography is being guided by agencies like ENISA (European Union Agency for Cybersecurity). ENISA has published reports and frameworks to help EU member states assess their cryptographic assets and prepare migration strategies. Unlike the centralized NIST model, the EU approach is more distributed, requiring coordination across national governments, private industries, and pan-European institutions. This model encourages local adaptation while aligning with international standards to ensure interoperability. Research projects under Horizon Europe and other EU funding programs have also promoted the development and testing of PQC in various sectors including banking, telecommunications, and government services.

Japan has adopted a balanced approach to cryptographic transition, emphasizing both research and practical deployment. The Ministry of Internal Affairs and Communications and the National Institute of Information and Communications Technology (NICT) have been central to these efforts.

Japan's strategy focuses on integrating PQC into critical infrastructure and promoting domestic cryptographic research. Collaborative efforts with universities and industry stakeholders have resulted in pilot projects to test PQC in secure communications and digital identity systems. Japan is also participating in international standardization bodies, ensuring that its national approach remains aligned with global trends.Across all these national efforts, common themes emerge. First is the emphasis on risk assessment and cryptographic inventory, which allows governments to identify where classical cryptography is currently in use and which systems are most vulnerable. Second is the prioritization of critical infrastructure, ensuring that systems involved in energy, defense, healthcare, and communications are secured first. Finally, international cooperation remains essential. Since cybersecurity is inherently global, alignment between countries on algorithm selection and implementation strategies is critical to avoid fragmentation and ensure cross-border compatibility.

11.5 ACADEMIC-INDUSTRY COLLABORATIONS IN PQC RESEARCH

The advancement of Post-Quantum Cryptography (PQC) depends heavily on robust collaborations between academia and industry. As quantum computing threatens to undermine traditional cryptographic systems, the urgency to develop, analyze, and deploy quantum-resistant algorithms has intensified. Academic institutions contribute theoretical foundations and algorithmic innovations, while industry partners provide real-world testing environments, implementation feedback, and large-scale deployment frameworks.

Academic researchers have long been at the forefront of cryptographic innovation. Universities and research institutes around the world have played a pivotal role in designing and analyzing PQC algorithms. Many of the candidate algorithms submitted to the National Institute of Standards and Technology (NIST) for PQC standardization originated from university-led teams or joint academic collaborations.

These teams leveraged deep expertise in number theory, lattice-based cryptography, code-based encryption, multivariate polynomial systems, and hash-based schemes. Their rigorous analysis, formal proofs, and vulnerability testing provided the initial validation required for these algorithms to be considered for global standardization.

At the same time, industry partners have brought critical perspectives to PQC research by testing how these algorithms perform under real-world conditions. Companies such as Google, IBM, Microsoft, Amazon, and Cloudflare have been instrumental in integrating PQC into experimental and operational systems.

They have evaluated the performance of PQC algorithms in applications like TLS (Transport Layer Security), VPNs, authentication systems, and cloud infrastructure. These industrial trials have identified practical challenges such as key size limitations, latency issues, and compatibility constraints, all of which feed back into academic refinement. For example, Google's CECPQ experiments combined NTRU or Kyber with traditional elliptic-curve algorithms in hybrid key exchanges, enabling performance benchmarking in real-time web traffic. Another important area of collaboration is the development of open-source libraries and testbeds. Joint efforts have produced libraries such as liboqs (Open Quantum Safe project), which provide implementations of PQC algorithms and support for integration with cryptographic protocols like OpenSSL. These libraries serve as platforms for both research and development, allowing students, scholars, engineers, and security professionals to experiment with and evaluate post-quantum algorithms. They also facilitate interoperability and consistency across different platforms, which is essential for broader adoption.

Funding mechanisms also play a central role in fostering academic-industry partnerships. Government agencies such as DARPA, the European Commission, and Japan's NICT have provided grants that specifically encourage joint research projects. These programs support interdisciplinary teams working on areas ranging from algorithm design and side-channel resistance to software optimization and standards development. They also ensure that small startups, large corporations, and research groups can collaborate under structured timelines and measurable goals.

CHAPTER

12 Appendices

12.1 GLOSSARY OF KEY TERMS

Algorithm Agility: The ability of a cryptographic system to transition smoothly between different cryptographic algorithms without requiring extensive redesign, ensuring adaptability to future security threats.

Asymmetric Cryptography: A cryptographic system that uses pairs of keys (public and private) for encryption and decryption, foundational for secure communications and digital signatures. Vulnerable to quantum attacks via Shor's algorithm.

Brute-Force Attack: An attack method where an adversary systematically checks all possible keys or inputs to break encryption. Grover's quantum algorithm can speed up brute-force searches, effectively halving the security strength of symmetric key systems.

Classical Cryptography: Traditional cryptographic methods (e.g., RSA, ECC, AES) that rely on mathematical problems solvable by classical computers but vulnerable to quantum attacks.

CRYSTALS (Cryptographic Suite for Algebraic Lattices): A family of NIST-selected post-quantum cryptographic algorithms, including Kyber (key encapsulation) and Dilithium (digital signatures), based on lattice-based cryptography.

Digital Signature: A cryptographic method for verifying the authenticity and integrity of digital messages or documents. Post-quantum alternatives include lattice-based (Dilithium), hash-based (SPHINCS+), and multivariate schemes.

Elliptic Curve Cryptography (ECC): A public-key cryptography system based on elliptic curve theory, widely used for key exchange and digital signatures. Susceptible to quantum attacks via Shor's algorithm.

Falcon: A NIST-standardized post-quantum digital signature algorithm based on lattice cryptography, designed for efficient verification but with complex key generation.

Grover's Algorithm: A quantum algorithm that provides a quadratic speedup for unstructured search problems, effectively reducing the security of symmetric key cryptography (e.g., AES-256 becomes as strong as AES-128 against quantum attacks).

Hash-Based Cryptography: A post-quantum approach relying on cryptographic hash functions (e.g., SPHINCS+ for signatures). Resistant to quantum attacks but often requires larger key sizes.

Hybrid Cryptography: A transitional approach combining classical and post-quantum algorithms to maintain security during migration (e.g., using both RSA and Kyber for key exchange).

Key Encapsulation Mechanism (KEM): A cryptographic protocol for securely exchanging symmetric keys. Post-quantum KEMs include Kyber (lattice-based) and Classic McEliece (code-based).

Kyber: A lattice-based KEM selected by NIST for standardization, offering efficient key exchange resistant to quantum attacks.

Lattice-Based Cryptography: A leading post-quantum approach relying on the hardness of lattice problems (e.g., Learning With Errors, or LWE). Includes algorithms like Kyber and Dilithium.

Migration Strategy: A planned approach for transitioning from classical to post-quantum cryptography, including cryptographic inventory, risk assessment, and phased implementation.

NIST PQC Standardization: A multi-year project by the U.S. National Institute of Standards and Technology (NIST) to evaluate and standardize quantum-resistant cryptographic algorithms.

Post-Quantum Cryptography (PQC): Cryptographic algorithms designed to be secure against attacks by both classical and quantum computers.

Quantum Computer: A computing device leveraging quantum-mechanical phenomena (e.g., superposition, entanglement) to perform calculations exponentially faster than classical computers for certain problems.

Quantum Resistance: The property of a cryptographic system remaining secure against attacks from quantum computers.

RSA Cryptography: A widely used public-key cryptosystem based on integer factorization, vulnerable to quantum attacks via Shor's algorithm.

Shor's Algorithm: A quantum algorithm capable of efficiently solving integer factorization and discrete logarithms, breaking RSA, ECC, and other classical public-key cryptosystems.

SPHINCS+: A stateless, hash-based digital signature scheme standardized by NIST as a post-quantum alternative.

Symmetric Cryptography: Encryption using a single shared key (e.g., AES). While quantum-resistant, Grover's algorithm reduces its effective security strength.

Threat Model: A structured representation of potential security risks, including quantum attacks, side-channel vulnerabilities, and implementation flaws in PQC systems.

XMSS (Extended Merkle Signature Scheme): A hash-based post-quantum signature scheme used in some blockchain applications.

Zero-Knowledge Proofs (ZKP): Cryptographic methods allowing one party to prove knowledge of a secret without revealing it. Some post-quantum variants are being explored for quantum-resistant authentication.

12.2 PQC ALGORITHM COMPARISON TABLES

Table: NIST-Selected PQC Algorithms (2022 Standardization)

Algorithm	Type	Security Claim	Key Size (approx.)	Signature Size (approx.)	Performance	Use Case
CRYSTALS-Kyber	Lattice-based KEM	L3 (AES-192 equiv.)	800-1,500 bytes	N/A	Fast encryption	General-purpose key exchange
CRYSTALS-Dilithium	Lattice-based signature	L3 (AES-192 equiv.)	1,300-2,500 bytes	2,400-4,500 bytes	Fast verification	Digital signatures
Falcon	Lattice-based signature	L5 (AES-256 equiv.)	900-1,300 bytes	600-1,300 bytes	Fast signing	Compact signatures (IoT, mobile)
SPHINCS+	Hash-based signature	L1-L5 (AES-128-256 equiv.)	1,000 bytes	8,000-50,000 bytes	Slow but simple	Backup/ stateless signatures

Table: Alternate PQC Candidates (NIST Round 4 Finalists)

Algorithm	Type	Security Claim	Key Size	Signature/ CT Size	Pros	Cons
Classic McEliece	Code-based KEM	L5 (AES-256 equiv.)	261,120 bytes	128-240 bytes	Longest-studied PQC	Huge public keys
BIKE	Code-based KEM	L1-L3	1,541-5,062 bytes	1,541-5,062 bytes	Smaller than McEliece	Slower than lattice
SIKE (Broken 2022)	Isogeny-based KEM	-	564 bytes	564 bytes	Ultra-compact	Fully broken by classical attack

Table: Performance Benchmark Comparison

Algorithm	KeyGen (ops/sec)	Encaps/ Sign (ops/sec)	Decaps/ Verify (ops/sec)	Memory Usage
Kyber-768	10,200	8,500	7,200	Low
Dilithium-3	1,150	850	2,100	Medium
Falcon-512	42	110	320	High (FFT req.)
SPHINCS+-128f	18	5	14	Very High

Table: Security Level Comparison

Security Level	Classical Equivalent	Matching PQC Algorithms
L1	128-bit (AES-128)	Kyber-512, Dilithium-2
L3	192-bit (AES-192)	Kyber-768, Dilithium-3
L5	256-bit (AES-256)	Kyber-1024, Falcon-1024

Table: Implementation Considerations

Factor	Lattice (Kyber/Dilithium)	Hash (SPHINCS+)	Code (McEliece)
Maturity	High (NIST selected)	Medium	Very High (since 1978)
Key Size	Medium	Small (pubkey)	Extremely Large
Side-Channel Risk	High (complex math)	Low	Medium
Hardware Friendliness	Good (needs FFT)	Excellent	Poor

Table: Recommended Use Cases

Application	Recommended Algorithms	Rationale
TLS/Web	Kyber + Dilithium	Balance of speed & security
IoT Devices	Falcon	Small signatures
Long-Term Archiving	SPHINCS+	No reliance on math assumptions
Government Comms	Kyber-1024 + Falcon-1024	Highest security tier

12.3 RECOMMENDED READING AND RESEARCH PAPERS

- **BIKE (Bit Flipping Key Encapsulation):** Aragon, N., et al. (2017). *"BIKE: Bit Flipping Key Encapsulation."*, A code-based KEM alternative to McEliece with smaller keys.

- **Classic McEliece:** McEliece, R. (1978). *"A Public-Key Cryptosystem Based on Algebraic Coding Theory."*, The original code-based encryption system still considered quantum-resistant.
- **CRYSTALS-Dilithium:** Ducas, L., et al. (2018). *"CRYSTALS - Dilithium: Digital Signatures from Module Lattices."*, Presents Dilithium, a lattice-based digital signature scheme.
- **CRYSTALS-Kyber:** Bos, J., et al. (2018). *"CRYSTALS - Kyber: A CCA-Secure Module-Lattice-Based KEM."*, Introduces Kyber, a lattice-based key encapsulation mechanism (KEM).
- **Falcon Signatures:** Fouque, P.-A., et al. (2018). *"Falcon: Fast-Fourier Lattice-Based Compact Signatures over NTRU."*, Describes Falcon, an efficient short-signature lattice scheme.
- **Fault Injection on PQC:** Espitau, T., et al. (2020). *"Fault Attacks on CRYSTALS-Dilithium."*, Explores how physical attacks can compromise lattice signatures.
- **Grover's Algorithm & Symmetric Security:** Grover, L. (1996). *"A Fast Quantum Mechanical Algorithm for Database Search."*, Established that quantum computers provide quadratic speedup for brute-force attacks.
- **Hybrid PQ/Classical Cryptography:** Bindel, N., et al. (2019). *"Transitioning to a Quantum-Resistant Public Key Infrastructure."*, Discusses strategies for combining PQC with traditional crypto.
- **Introduction to Post-Quantum Cryptography (NIST, 2020):** Chen, L., et al., A high-level overview of PQC principles and NIST's standardization.
- **NIST PQC Project Updates:** Follow NIST's PQC announcements for future algorithm revisions.
- **NIST PQC Standardization Project:** *"Post-Quantum Cryptography Standardization"* (NIST, 2016-2022), The official NIST documentation on selected algorithms (Kyber, Dilithium, Falcon, SPHINCS+) and their security evaluations.
- **Post-Quantum Cryptography (Springer, 2009):** Bernstein, D., Buchmann, J., Dahmen, E. (Eds.), Early but foundational book covering multiple PQC approaches.
- **PQC Side-Channel Risks:** Primas, R., et al. (2019). *"Single-Trace Side-Channel Attacks on Masked Lattice-Based Encryption."*, Demonstrates side-channel vulnerabilities in lattice-based PQC.

- **Shor's Algorithm & Quantum Threat:** Shor, P. (1994). *"Algorithms for Quantum Computation: Discrete Logarithms and Factoring."*, The seminal paper proving that quantum computers can break RSA and ECC.
- **SIKE Cryptanalysis (2022 Break):** Castryck, W., Decru, T. (2022). *"An Efficient Key Recovery Attack on SIDH."*, The paper that broke SIKE using a classical attack.
- **SPHINCS+ Stateless Signatures:** Bernstein, D., et al. (2019). *"SPHINCS+: Stateless Hash-Based Signatures with Chosen Parameters."*, A quantum-resistant signature scheme without stateful requirements.
- **Supersingular Isogeny Diffie-Hellman (SIDH):** De Feo, L., et al. (2011). *"Towards Quantum-Resistant Cryptosystems from Supersingular Elliptic Curve Isogenies."*, Introduced isogeny-based cryptography before SIKE's 2022 break.
- **TLS 1.3 with Post-Quantum KEMs:** Stebila, D., Mosca, M. (2018). *"Post-Quantum Key Exchange for the Internet."*, Examines integrating PQC into TLS for secure web traffic.
- **XMSS & LMS Hash-Based Signatures:** Buchmann, J., et al. (2011). *"On the Security of the Winternitz One-Time Signature Scheme."*, Early work on hash-based signatures adopted in NIST standards.

12.4 PQC CHALLENGE SETS AND BENCHMARK TOOLS

12.4.1 NIST PQC Benchmarking Framework

- **Purpose**: Official reference implementations for testing NIST-selected algorithms (Kyber, Dilithium, Falcon, SPHINCS+)
- Components:
  - Reference C code for all finalists
  - Known-answer tests (KATs) for validation
  - Performance measurement suites
- **Access**: NIST PQC Submission Packages

12.4.2 PQCRYPTO Project Challenges

- EU Horizon 2020 Initiative
  - Provides standardized attack scenarios for:
    - Lattice reduction challenges
    - Code decoding problems
    - Hash collision resistance tests
- **Includes**: Concrete parameter sets matching NIST security levels
- **Repository**: PQCRYPTO GitHub

12.4.3 Performance Benchmarking Suites

Open Quantum Safe (OQS) Benchmarking Toolkit

- Features:
  - Cross-platform performance testing (x86, ARM, RISC-V)
  - Memory/cycle-accurate measurements
  - TLS 1.3 integration benchmarks
- **Supported Algorithms:** All NIST finalists + alternate candidates
- **Tool:** `speed_kem` and `speed_sig` test binaries
- **GitHub:** open-quantum-safe/liboqs

PQClean Implementation Benchmarking

- **Focus:** Pure-C implementations for fair comparison
  - Excludes platform-specific optimizations
  - Measures algorithmic overhead only
- **Metrics:** Clock cycles per operation, stack usage
- **Data:** PQClean Benchmark Reports

12.4.4 Cryptanalysis Challenge Sets

LWE Challenge Database

- **Hosted:** University of Cincinnati
  - Graded instances from easy (n=40) to hard (n=160)
  - Tracks best-known attacks (BKZ, sieving)
- **Purpose:** Evaluate concrete hardness of lattice problems
- **Access:** LWE Challenge Portal

Code-Based Cryptography Challenges

- **McEliece Attack Challenges:**
  - Classic McEliece parameter sets
  - Decoding challenges for BIKE/HQC
- **Maintained:** by TU Eindhoven PQC Lab
- **Submission:** Via IACR ePrint

12.4.5 Hardware Benchmarking Tools

PQAX: Post-Quantum ASIC Benchmark Suite

- **Developed:** ETH Zurich Secure Hardware Lab
- **Capabilities:**

- ASIC/FPGA area measurements
- Power/energy profiling
- Side-channel leakage evaluation

- **Algorithms:** Optimized Verilog/VHDL implementations
- **Publication:** CHES 2022 Proceedings

ARM Cortex-M PQC Benchmarking

- **Target:** IoT devices (STM32, nRF52)
  - Real-time performance metrics
  - Flash/RAM usage statistics
- **Toolchain:** Keil MDK with PQC patches
- **Reference:** ARM Mbed TLS PQC Port

12.4.6 Cloud-Based Benchmarking Platforms

AWS Quantum Computing Benchmark Service

- **Features:**
  - Parallel testing across instance types
  - Cost-per-operation calculations
  - Hybrid classical/quantum simulation
- **Integrated Algorithms:** All NIST PQC finalists
- **Access:** AWS Marketplace (PQC Benchmarking AMI)

Google PQC Benchmarking Dashboard

- **Capabilities:**
  - WebAssembly-based in-browser testing
  - Historical performance tracking
  - Algorithm comparison visualizations
- **URL:** pqbench.google.dev

12.4.7 Specialized Challenge Tracks

Side-Channel Attack Challenges

- **DASCA PQC Leakage Database:**
  - Power/EM traces for masked implementations
  - Fault injection test vectors
- **Organizer:** Ruhr University Bochum
- **Dataset:** SCA-DB.org

Quantum Hybrid Attack Simulation

- **IBM Qiskit PQC Challenges:**
  - Simulates quantum-classical hybrid attacks
  - Graded difficulty levels
- **Platform:** IBM Quantum Experience
- **Notebooks:** Qiskit PQC Toolkit

12.4.8 Educational Challenge Sets

CryptoHack PQC Track

- **Format:** CTF-style challenges
  - LWE solving
  - Signature forgery
  - Parameter recovery
- **Difficulty:** Beginner to advanced
- **Link:** CryptoHack.org/PQC

PQCrypto School Challenges

- **Annual training program:**
  - Algorithm identification
  - Implementation flaws
  - Optimization puzzles
- **Archive:** PQCrypto.org/school

12.4.9 Usage Recommendations

- **For Standard Compliance Testing:** Use NIST KATs + OQS benchmarking
- **For Cryptanalysis Research:** LWE Challenge + PQCRYPTO datasets
- **For Hardware Design:** PQAX + ARM Mbed TLS results
- **For Education:** CryptoHack + Qiskit notebooks

12.5 LIST OF PQC TOOLS, LIBRARIES, AND STANDARDS

12.5.1 NIST-Standardized PQC Algorithms

- **CRYSTALS-Kyber** (Key Encapsulation)
- **CRYSTALS-Dilithium** (Digital Signatures)
- **Falcon** (Compact Digital Signatures)
- **SPHINCS+** (Stateless Hash-Based Signatures)

12.5.2 Open-Source PQC Libraries

Library	Language	Features
Open Quantum Safe (OQS)	*C/Python*	*Reference implementations of all NIST PQC finalists + TLS integration*
PQClean	*C/ASM*	*Clean-room implementations for benchmarking*
libpqcrypto	*C*	*Optimized for embedded systems*
Bouncy Castle PQC	*Java*	*NIST algorithms in Java Cryptography Architecture (JCA)*
WolfSSL PQC	*C*	*Commercial-grade TLS with PQC support*

12.5.3 PQC Protocol Integrations

- **OQS-OpenSSL** (GitHub): Quantum-safe TLS 1.3 fork.
- **OQS-BoringSSL** (GitHub): Google's BoringSSL + PQC.
- **PQCrypto-SIDH** (GitHub): *Deprecated* (SIKE was broken in 2022).
- **ProtonMail PGP-PQC** (Docs): Experimental PQ/PGP hybrid encryption.

12.5.4 Hardware Acceleration Tools

Tool	Purpose
PQAX (ETH Zurich)	ASIC/FPGA benchmarks for PQC
ARM Mbed TLS PQC (GitHub)	ARM-optimized PQC for IoT
Intel IPP-Crypto PQC (Intel Docs)	x86-64 optimized lattice crypto

12.5.5 Side-Channel Analysis Tools

- **ELMO** (GitHub): Leakage assessment for PQC.
- **Dudect** (GitHub): Constant-time execution testing.
- **ChipWhisperer** (NewAE): Power/EM analysis for PQC implementations.

12.5.6 Cryptographic Standards & APIs

- **NIST SP 800-208** (PDF): PQC migration guidelines.
- **IETF Drafts:**
  - draft-ietf-tls-hybrid-design: Hybrid PQ/classical TLS.
  - draft-ounsworth-pq-composite-sigs: Composite signatures.
- **FIPS 203/204/205** (Upcoming): NIST's formal PQC standards (2024-25).

12.5.7 Educational & Testing Tools

Tool	Purpose
PQChallenges (GitHub)	CTF-style PQC cryptanalysis
Qiskit PQC Module (IBM)	Quantum attack simulations
CryptHack PQC Track (Link)	Learning via challenges

12.5.8 Commercial PQC Solutions

Vendor	Product
Google	PQC in Chrome
Cloudflare	PQC TLS 1.3
AWS	KMS PQC Support
Thales	PQC HSMs

References

Algebraic Codes for PQC: Judy Walker AMS, 2019

Banking on PQC: Anthony Glees, SWIFT Risk Books, 2022

Beyond NIST PQC: Next-Gen Algorithms: Vadim Lyubashevsky Springer, 2024

Code-Based Cryptography: Raphael Overbeck, Nicolas Sendrier Springer, 2009

Computational Number Theory for PQC: Richard Brent Cambridge University Press, 2022

CRYSTALS-Kyber and Dilithium: A Technical Deep Dive: Peter Schwabe, Bas Westerbaan IACR, 2022

Enterprise PQC Migration: Tim Hollebeek, DigiCert Wiley, 2023

Ethics of Quantum-Safe Security: Bruce Schneier W. W. Norton, 2024

Falcon: Fast-Fourier Lattice Signatures: Pierre-Alain Fouque et al. Springer, 2020

Fault Injection in PQC: Amir Moradi Springer, 2022

Grover's Algorithm in Cryptanalysis: Nicolas Gisin World Scientific, 2021

Guide to NIST Post-Quantum Cryptography: Lily Chen, Dustin Moody, NIST NIST SP 1800-208, 2023

Hardware Acceleration for PQC: Ingrid Verbauwhede, Tim Güneysu Springer, 2022

Hardware Trojans in PQC: Markus G. Kuhn CRC Press, 2023

Hybrid Cryptography: Classical + PQC: Tanja Lange, Daniel J. Bernstein Springer, 2020

Introduction to Post-Quantum Cryptography: Jonathan Katz Chapman & Hall, 2023

Isogenies in Cryptography: Luca De Feo SIAM, 2021

Lattice-Based Cryptography: Chris Peikert MIT Press, 2016

Lattices in Cryptography: Oded Regev Princeton University Press, 2023

Masking PQC Implementations: Sonia Belaïd, Jean-Sébastien Coron IACR Books, 2021

Mathematics of Public Key Cryptography: Steven D. Galbraith Cambridge University Press, 2012

McEliece Cryptosystem: 40 Years of Security: Rafael Misoczki, Nicolas Sendrier Now Publishers, 2018

Multivariate Cryptography: Jintai Ding, Bo-Yin Yang Springer, 2020

NIST PQC Standardization Process: Dustin Moody NIST, 2022

Open Quantum Safe: Building PQC Libraries: Douglas Stebila, Michele Mosca O'Reilly, 2022

Physical Security for PQC: Pankaj Rohatgi Wiley, 2022

Post-Quantum Cryptography: Daniel J. Bernstein, Johannes Buchmann, Erik Dahmen, Eds. Springer, 2009

PQC and AI Security Shafi Goldwasser Now Publishers, 2023

PQC at Google: A Case Study: Matt Braithwaite Google Research, 2023

PQC for Embedded Systems: Patrick Longa, Wai-Kong Lee CRC Press, 2023

PQC for Engineers: Jean-Philippe Aumasson No Starch Press, 2023

PQC for Governments: Whitfield Diffie, Susan Landau MIT Press, 2021

PQC for IoT: Javier Lopez, Rolf Blom Springer, 2021

PQC in 5G Networks: David Wong, Meta Artech House, 2023

PQC in Cloud Security: Victor Chang, AWS Packt, 2023

PQC in Python: Samuel J. Lomonaco Jr. Manning, 2023

PQC in Space Communications: Scott Aaronson NASA Press, 2022

PQC in TLS and PKI: Eric Rescorla, Richard Barnes Addison-Wesley, 2022

PQC: A Graduate Course: Chris Peikert, Vinod Vaikuntanathan MIT OpenCourseWare, 2023

Practical Post-Quantum Cryptography in C: Michael Scott Apress, 2021

Quantum Attacks on Classical Cryptography: Mark Ettinger, Peter Høyer Wiley, 2019

Quantum Computing and Cryptography: Michele Mosca CRC Press, 2018

Quantum Cryptography Beyond QKD: Artur Ekert Cambridge University Press, 2022

Quantum Internet and PQC: Stephanie Wehner Cambridge University Press, 2023

Quantum-Resistant Blockchain: Andreas Antonopoulos Merkle Bloom, 2023

Quantum-Safe Cryptography Algorithms: Raymond T. Ng World Scientific, 2022

Shor's Algorithm and Its Impact: Michele Mosca, Philip Kaye Oxford University Press, 2020

Side-Channel Attacks on PQC: Colin O'Flynn Newnes, 2023

SPHINCS+ and Hash-Based Signatures: Andreas Hülsing et al. ACM Books, 2021

The PQC Handbook: Daniel J. Bernstein, Tanja Lange De Gruyter, 2024

Index

www.ingramcontent.com/pod-product-compliance
Lightning Source LLC
LaVergne TN
LVHW081938200926
848484LV00009B/2465
* 9 7 8 1 7 8 7 1 5 6 2 4 1 *